ASME ENGINEER'S DATA BOOK

by Clifford Matthews

SME Press . . . New York . . . 2001

Library of Congress Cataloging-in-Publication Data

Matthews, Clifford.

ASME engineer's data book / by Clifford Matthews.

p. cm.

Includes bibliographical references.

ISBN 0-7918-0155-1 (pbk.)

1. Mechanical Engineeering—Handbooks, manuals, etc.
I. Title.

TJ157 .M392 2001

621—dc21 00-054279

Contents

Preface

The objective of this ASME mechanical engineers' data book is to provide a concise and useful source of up-to-date information for the student or practicing Professional Engineer. Despite the proliferation of specialized information sources, there is still a need for basic data on established engineering rules, conventions, and modern developments to be available in an easily assimilated format.

An engineer cannot afford to ignore the importance of engineering data and rules. Basic theoretical principles underlie the design of all the hardware of engineering. The practical processes of design, material choice, tolerances, joining, and testing form the foundation of its manufacture and operation.

Technical standards are also important. Standards represent accumulated knowledge and form invaluable guidelines for the design and manufacture of material and equipment. For this reason you will find extensive references to technical standards in this book.

The purpose of the book is to provide a basic set of engineering data that you will find useful. It is divided into 22 sections, each containing specific "discipline" information. Units and conversions are covered in Section 2—Metric and SI units are used in many industries around the world. Section 22 contains lists of Web sites from major American engineering associations and standards organizations: Web sites are currently one of the fastest-growing ways of finding engineering-standards-based information.

You will see that various pages in the book contain "quick guidelines" and "rules of thumb." Don't expect these all to have robust theoretical backing—they are included simply because I have found that they *work*. I have tried to make this book a practical source of mechanical engineering information that you can use in the day-to-day activities of an engineering career. This means it belongs in your pocket, not in the bottom drawer of your desk.

Finally, it is important that the content of this data book continues to reflect the information needed and used by student and experienced engineers. If you have any suggestions for future content (or comment on the existing content) please send them to me at the following e-mail address:

Usdatabook2000@aol.com

Cliff Matthews

Foreword from ASME Press

The intention of this Data Book is to provide a quick and handy reference for mechanical engineers. Due to the pocket size of the book, it was impossible to provide comprehensive coverage of all areas of mechanical engineering. In consultation with a five-member editorial review board, we provided the author with some guidelines for what we collectively felt was the most useful information to include.

It also helped that this book was derived from a very successful volume written by the author for the Institution of Mechanical Engineers (IMechE) in the U.K. Of course, the book had to be substantially modified for American engineers. We would like to thank the staff at Professional Engineering Publishing for making this ASME Press edition possible.

ASME Press is a book publishing imprint of the American Society of Mechanical Engineers (ASME International), which was founded in 1880 and currently has over 125,000 members. We hope that with this book, we will help to fulfill some of the intended purposes of ASME, which according to the Society's Bylaws are:

- To "promote the art, science, and practice of mechanical engineering and the allied arts and sciences."
- To "encourage original research; foster engineering education; advance the standards of engineering; promote the exchange of information among engineers and others; broaden the usefulness of the engineering profession in cooperation with other engineering and technical societies; and promote a high level of ethical practice."
- To fulfill its purposes, ASME provides many services to its members and to the engineering community in general. These services can be grouped into the following categories:

Dissemination of Knowledge: ASME publishes books, journals, and conference proceedings to disseminate knowledge and experience of value to engineers. ASME also sponsors technical conferences for the presentation and discussion of original papers from engineers around the world.

Codes & Standards: ASME is well-known for providing technical standards, codes, formulae, and recommended practices, which help ensure the quality and safety of many engineered products.

Education: ASME offers numerous continuing education and professional development courses in cities all around the country. ASME also sponsors programs aimed at promoting the engineering profession among college and high school students.

Professional Standards, Usefulness of the Profession, and Ethical Practice: ASME offers awards and other honors to recognize meritorious contributions to engineering and to encourage contributions to the field. ASME supports the adoption of a high standard of attainment for the granting of the legal right to practice professional engineering.

The Society strongly encourages the personal and professional development of young engineers. The Society also strives to promote the engineering profession to the community at large by publicizing the achievements of engineers and by encouraging their participation in public affairs. Lastly, ASME maintains a Code of Ethics of Engineers consistent with the high standards of the profession.

(More information about ASME can be found on the Society's web site, at www.asme.org.)

To varying degrees, this Data Book will contribute to all of the Society's goals, but it will contribute particularly in the area of disseminating useful knowledge to engineers. This book will be especially helpful to young engineers as a quick source of basic information on a broad range of engineering topics. It will also guide readers to more comprehensive sources of information in specific areas.

We welcome any suggestions from readers as to how we may make future editions of the Data Book even more helpful.

Finally, we would like to thank the members of the editorial review board for their help and input:

Eugene A. Avallone, PE (Life Member, ASME)
Omer W. Blodgett, DSc, PE (Life Fellow, ASME)
Steve R. Daniewicz (Associate Member, ASME)
Franklin Fisher, PhD, PE (Life Member, ASME)
Jay Lee, DSc (Member, ASME)

Editor, ASME Press

Introduction

The Role of Technical Standards

What role do published technical standards play in mechanical engineering? Standards have been part of the engineering scene in the United States since the early days of manufacturing when they were introduced to try to solve the problem of substandard products. In those early days they were influential in increasing the availability (and reducing the price) of basic iron and steel products.

What has happened since then? Standards bodies around the world have proliferated, working more or less independently, but all subject to the same engineering laws and practical constraints. They have developed slightly different ways of looking at technical problems, which is not such a bad thing—the engineering world would be less of an interesting place if everyone saw things in precisely the same way. Varied though they may be, published standards represent good practice. Their ideas are tried and tested, and they operate across the spectrum of engineering practice, from design and manufacture to testing and operation.

Standards also exert influence on the commercial practices of U.S. companies. The QS–9000 and ISO 9000 quality management standards are now almost universally accepted as *the* QA model to follow. More and more companies in the U.S. are becoming certified—a trend that can only improve engineering management practices worldwide.

Technical standards are an important model for technical conformity in all fields. They affect just about every mechanical engineering product from pipelines to paper clips. From the practical viewpoint it is worth considering that, without standards, the design and manufacture of even the most basic engineering design would have to be started from the beginning.

Section 1

Essential Engineering Mathematics

Powers and Roots

$$a^n.a^m = a^{n+m} \qquad \frac{a^n}{a^m} = a^{n-m} \qquad ab^n = a^n b^n \qquad \left(\frac{a}{b}\right)^n = \frac{a^n}{b^n}$$

$$(a^n)^m = (a^m)^n = a^{nm} \qquad (\sqrt[n]{a})^n = a \qquad a^{1/n} = \sqrt[n]{a}$$

$$a^{n/m} = \sqrt[m]{a^n} \qquad \sqrt[n]{ab} = \sqrt[n]{a}.\sqrt[n]{b}$$

Logarithms

$$\log_a a = 1 \qquad \log_a 1 = 0 \qquad (\log_a M)\, N = \log_a M + \log_a N$$

$$\log_b N = \frac{\log_a N}{\log_a b} \qquad \log_b b^N = N \qquad b^{\log_b N} = N$$

The Quadratic Equation

A quadratic equation is one in the form $ax^2 + bx + c = 0$
Where a, b, and c are constants.

The solution is: $x = \dfrac{-b \pm \sqrt{b^2 - 4ac}}{2a}$

Trigonometric Functions

$$\sin\alpha = \frac{y}{r}$$

$$\cos\alpha = \frac{x}{r}$$

$$\tan\alpha = \frac{y}{x}$$

$$\cot\alpha = \frac{x}{y}$$

Fig. 1–1

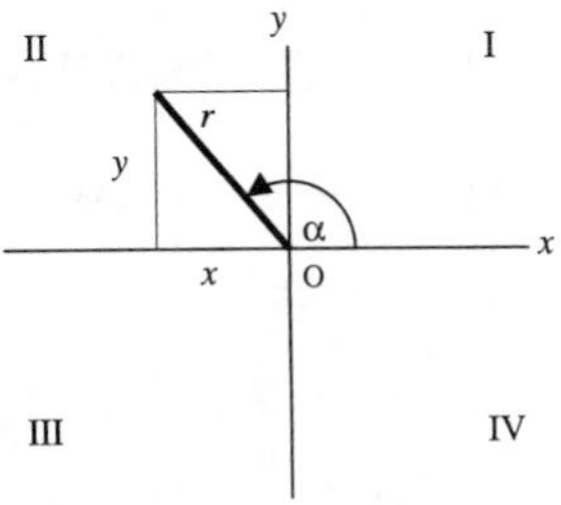

$$\sec \alpha = \frac{r}{x}$$

$$\text{cosec } \alpha = \frac{r}{y}$$

The signs of these functions depend on which quadrant they are in:

Quadrant	*Sin*	*Cos*	*Tan*	*Cot*	*Sec*	*Cosec*
I	+	+	+	+	+	+
II	+	−	−	−	−	+
III	−	−	+	+	−	−
IV	−	+	−	−	+	−

Trig Functions of Common Angles

	0°	*30°*	*45°*	*60°*	*90°*
Sin	0	$\frac{1}{2}$	$\sqrt{2}/2$	$\sqrt{3}/2$	1
Cos	1	$\sqrt{3}/2$	$\sqrt{2}/2$	$\frac{1}{2}$	0
Tan	0	$\sqrt{3}/3$	1	$\sqrt{3}$	∞
Cot	∞	$\sqrt{3}$	1	$\sqrt{3}/3$	0
Sec	1	$2\sqrt{3}/3$	$\sqrt{2}$	2	∞
Cosec	∞	2	$\sqrt{2}$	$2\sqrt{3}/3$	1

$$\sin \alpha = \frac{1}{\text{cosec } \alpha} \quad \cos \alpha = \frac{1}{\sec \alpha} \quad \tan \alpha = \frac{1}{\cot \alpha} = \frac{\sin \alpha}{\cos \alpha}$$

$$\sin^2\alpha + \cos^2\alpha = 1 \quad \sec^2\alpha - \tan^2\alpha = 1$$

$$\text{cosec}^2\alpha - \cot^2\alpha = 1$$

Differential Calculus

Derivatives	*Integrals*
$\frac{d}{dx}(u \pm v + \ldots) = \frac{du}{dx} \pm \frac{dv}{dx} \pm \ldots$	$\int d\,f(x) = f(x) + C$
$\frac{d}{dx}(uv) = \frac{udv}{dx} + \frac{vdu}{dx}$	$d \int f(x)\,dx = f(x)\,dx$
$\frac{d}{dx}\left(\frac{u}{v}\right) = v^2\left(\frac{vdu}{dx} - \frac{udv}{dx}\right)$	$\int af\,x\,dx = a \int f\,x\,dx$ (a = constant)
$\frac{d}{dx}(u^n) = nu^{n-1}\frac{du}{dx}$	$\int udv = uv - \int vdu$
$\frac{d}{dx}(\ln u) = \frac{1}{u}\frac{du}{dx}$	$\int u^n du = \frac{u^{n+1}}{n+1} + C \quad n \mp -1$
$\frac{d}{dx}(\tan u) = \sec^2 u \frac{du}{dx}$	$\int \frac{du}{u} = \ln u + C$
$\frac{d}{dx}(\sin u) = \cos u \frac{du}{dx}$	$\int e^u du = e^u + C$
$\frac{d}{dx}(\cos u) = -\sin u \frac{du}{dx}$	$\int \sin u\,du = -\cos u + C$
	$\int \cos u\,du = \sin u + C$
	$\int \tan u\,du = -\ln \cos u + C$

Section 2

Units

2.1 The Greek Alphabet

The Greek alphabet is used extensively in the United States to denote engineering quantities (see Table 2.1). Each letter can have various meanings, depending on the context in which it is used

Table 2.1 The Greek Alphabet

Name	*Symbol*	
	Capital	*Lower case*
alpha	A	α
beta	B	β
gamma	Γ	γ
delta	Δ	δ
epsilon	E	ϵ
zeta	Z	ζ
eta	H	η
theta	Θ	θ
iota	I	ι
kappa	K	κ
lambda	Λ	λ
mu	M	μ
nu	N	ν
xi	Ξ	ξ
omicron	O	o
pi	Π	π
rho	P	ρ
sigma	Σ	σ
tau	T	τ
upsilon	Υ	υ
phi	Φ	ϕ
chi	X	χ
psi	Ψ	ψ
omega	Ω	ω

2.2 Units Systems

The most commonly used system of units in the United States is the United States Customary System (USCS). The "MKS system" is a metric system still used in some European countries but gradually being superseded by the expanded Système International (SI) system.

2.2.1 The USCS System

Countries outside the United States often refer to this as the "inch-pound" system. The base units are:

Length:	foot (ft) = 12 inches (in.)
Force:	pound force (lbf)
Time:	second (s)
Temperature:	degrees Farenheit (°F)

2.2.2 The SI System

The strength of the SI system is its *coherence*. There are four mechanical and two electrical base units, from which all other quantities are derived. The mechanical ones are:

Length:	meter (m)
Mass:	kilogram (kg)
Time:	second (s)
Temperature:	Kelvin (K), or more commonly, degrees Centigrade (°C)

Other units are derived from these; e.g., the Newton (N) is defined as $N = kgm/s^2$. Formal SI conversion factors are listed in ASTM Standard E380.

2.2.3 SI Prefixes

As a rule, prefixes are applied to the basic SI unit, except for weight, where the prefix is used with the unit gram (g), not the basic SI unit kilogram (kg). Prefixes are not used for units of angular measurement (degrees, radians), time (seconds) or temperature (°C or K).

Prefixes should be chosen in such a way that the numerical value of a unit lies between 0.1 and 1000 (see Table 2.2)

eg:	28 kN	rather than	2.8×10^4N
	1.25 mm	rather than	0.00125 m
	9.3 kPa	rather than	9300 Pa

Table 2.2 SI Unit Prefixes

Multiplication Factor	*Prefix*	*Symbol*
1 000 000 000 000 000 000 000 000 = 10^{24}	yotta	Y
1 000 000 000 000 000 000 000 = 10^{21}	zetta	Z
1 000 000 000 000 000 000 = 10^{18}	exa	E
1 000 000 000 000 000 = 10^{15}	peta	P
1 000 000 000 000 = 10^{12}	tera	T
1 000 000 000 = 10^{9}	giga	G
1 000 000 = 10^{6}	mega	M
1 000 = 10^{3}	kilo	k
100 = 10^{2}	hicto	h
10 = 10^{1}	deka	da
0.1 = 10^{-1}	deci	d
0.01 = 10^{-2}	centi	c
0.001 = 10^{-3}	milli	m
0.000 001 = 10^{-6}	micro	μ
0.000 000 001 = 10^{-9}	nano	n
0.000 000 000 001 = 10^{-12}	pico	p
0.000 000 000 000 001 = 10^{-15}	femto	f
0.000 000 000 000 000 001 = 10^{-18}	atto	a
0.000 000 000 000 000 000 001 = 10^{-21}	zepto	z
0.000 000 000 000 000000 000 001 = 10^{-24}	yocto	y

2.3 Conversions

Units often need to be converted. The least confusing way to do this is by expressing *equality*:

For example: to convert 600 lb to kilgrams (kg)
Using 1kg = 2.205lb

Add denominators as

$$\frac{1kg}{\text{x}} = \frac{2.205lb}{600lb}$$

Solve for x

$$x = \frac{600 \times 1}{2.205} = 272.1$$

Hence 600 lb = 272.1kg

Setting out calculations in this way can help avoid confusion, particularly when they involve large numbers and/or several sequential stages of conversion.

2.3.1 Force

The USCS unit is the *pound force (lbf)*. Note that pound is also ambiguously used as a unit of mass (see Table 2.3).

Table 2.3 Force (F)

Unit	*lbf*	*gf*	*kgf*	*N*
1 pound (lbf)	1	453.6	0.4536	4.448
1 gram force (gf)	2.205×10^{-3}	1	0.001	9.807×10^{-3}
1 kilogram-force (kgf)	2.205	1000	1	9.807
1 Newton (N)	0.2248	102.0	0.1020	1

Note: Strictly, all the units in the table except the Newton (N) represent weight equivalents of mass, and so depend on the "standard" acceleration due to gravity (g). The true SI unit of force is the Newton (N), which is equivalent to 1 kgm/s^2.

2.3.2 Weight

The true weight of a body is a measure of the gravitational attraction of the earth on it. Since this attraction is a force, the weight of a body is correctly expressed in pounds force (lbf). See Fig. 2.1.

Mass is measured in pounds mass (lbm) or simply (lb)
Force (lbf) = mass (lbm) × g(ft/s^2)

Fig. 2–1

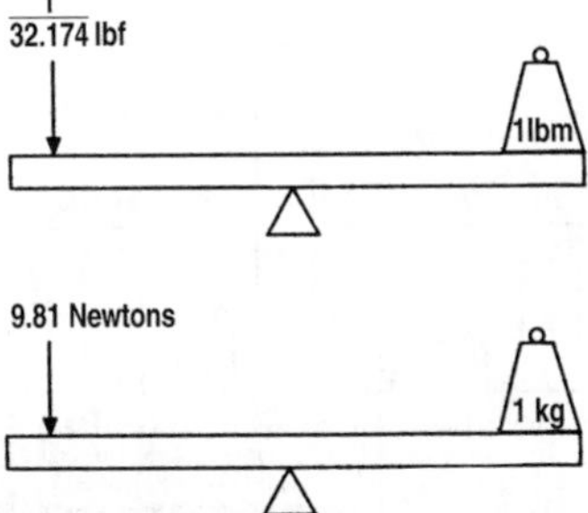

Or, in SI units: Force (N) = mass (kg) × g (m/s^2)
1 ton (US) = 2000 lb = 907.2 kg
1 tonne (metric) = 1000 kg = 2205 lb

2.3.3 Density

Density is defined as mass per unit volume. Table 2.4 shows the conversions between units.

Table 2.4 Density (ρ)

Unit	*lb/in³*	*lb/ft³*	*kg/m³*	*g/cm³*
1 lb per in³	1	1728	2.768×10^4	27.68
1 lb per ft³	5.787×10^{-4}	1	16.02	1.602×10^{-2}
1 kg per m³	3.613×10^{-5}	6.243×10^{-2}	1	0.001
1g per cm³	3613×10^{-2}	62.43	1000	1

2.3.4 Pressure

The base unit is the lbf/in² (or "psi")

1 Pa = 1 N/m²
1 Pa = 1.45038×10^{-4} lbf/in²

In practice, pressures in SI units are measured in MPa, bar, atmospheres, torr or the height of a liquid column, depending on the application. See Figs. 2.2, 2.3 and Table 2.5.

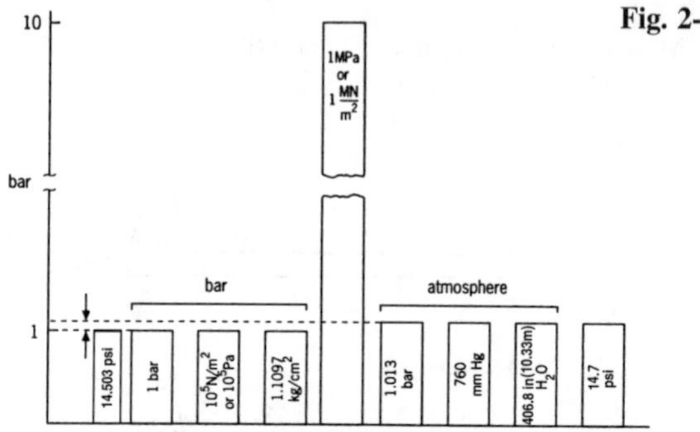

Fig. 2-2

Table 2.5 Pressure (ρ)

Unit	*lb/in² (psi)*	*lb/ft²*	*atm*	*in.H_2O*	*cm Hg*	*N/m² (Pa)*
1 lb per in^2 (psi)	1	144	6.805×10^{-2}	27.68	5 .171	6.895×10^{3}
1 lb per ft^2	6.944×10^{-3}	1	4.725×10^{-4}	0.1922	3.591×10^{-2}	47.88
1 atmosphere (atm)	14.70	2116	1	406.8	76	1.013×10^{5}
1 in of water at 39.2°F (4°C)	3.613×10^{-2}	5.02	2.458×10^{-3}	1	0.1868	249.1
1 cm of mercury at 32°F (0°C)	0.1934	27.85	1.316×10^{-2}	5.353	1	1333
1 N per m^2 (Pa)	1.450×10^{-4}	2.089×10^{-2}	9.869×10^{-6}	4.015×10^{-5}	7.501×10^{-4}	1

Fig. 2–3

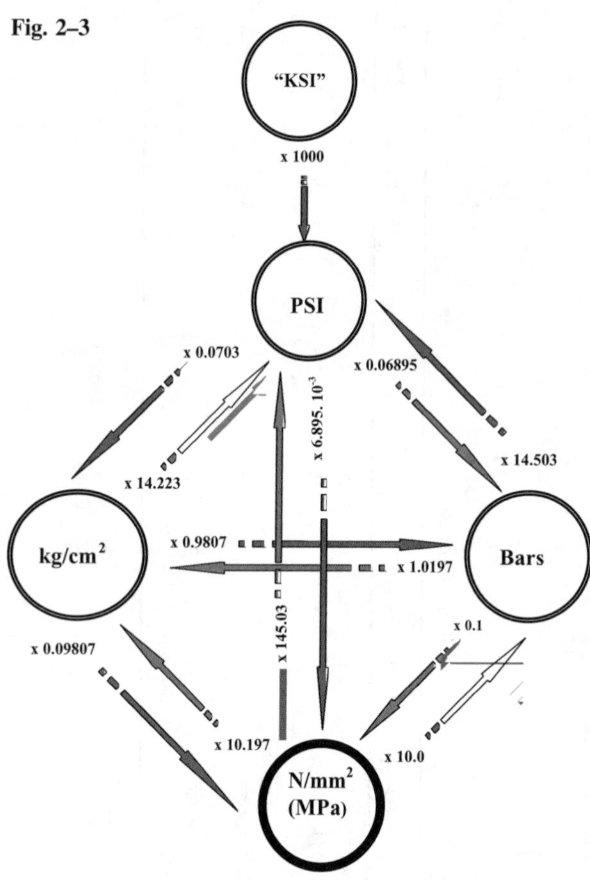

So for liquid columns:

1 in H_2O = 25.4 mm H_2O = 249.089 Pa
1 in Hg = 13.59 in H_2O = 3385.12 Pa = 33.85 mbar
1 mm Hg = 13.59 mm H_2O = 133.3224 Pa = 1.333224 mbar
1 mm H_2O = 9.80665 Pa
1 torr = 133.3224 Pa

For conversion of liquid column pressures, 1 in = 25.4 mm

2.3.5 Temperature

The basic unit of temperature is degrees Fahrenheit (°F). The SI unit is Kelvin (K). The most commonly used unit is degrees Celsius (°C).

Absolute zero is defined as 0 K or −273.15 °C, the point at which a perfect gas has zero volume. See Figures 2.4 and 2.5.

°C = 5/9 (°F − 32)
°F = 9/5 (°C) + 32

Fig. 2–4

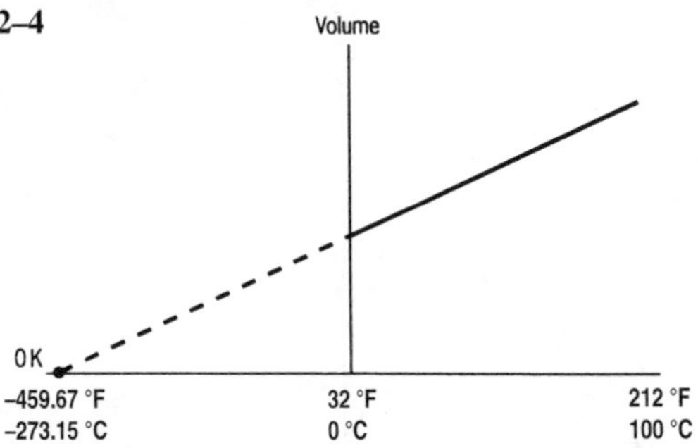

2.3.6 Heat and Work

The basic unit for heat "energy" is the British Thermal Unit (Btu).

Specific heat "energy" is measured in Btu/lb [in SI it is Joules per kilogram (J/kg)].

1 J/kg = 0.429923×10^{-3} Btu/lb

Table 2.6 shows common conversions

Specific heat is measured in Btu/lb°F [or in SI, Joules per kilogram Kelvin (J/kg K)].

1 Btu/lb °F = 4186.798 J/kgK.
1 J/kg K = 0.238846 10^{-3} Btu/lb °F
1 kcal/kg K = 4186.8 J/kg K

Heat flowrate is also defined as power, with the unit of Btu/h [or in SI in Watts (W)].

1 Btu/h = 0.07 cal/s = 0.293 W
1 W = 3.41214 Btu/h = 0.238846 cal/s

Fig. 2–5

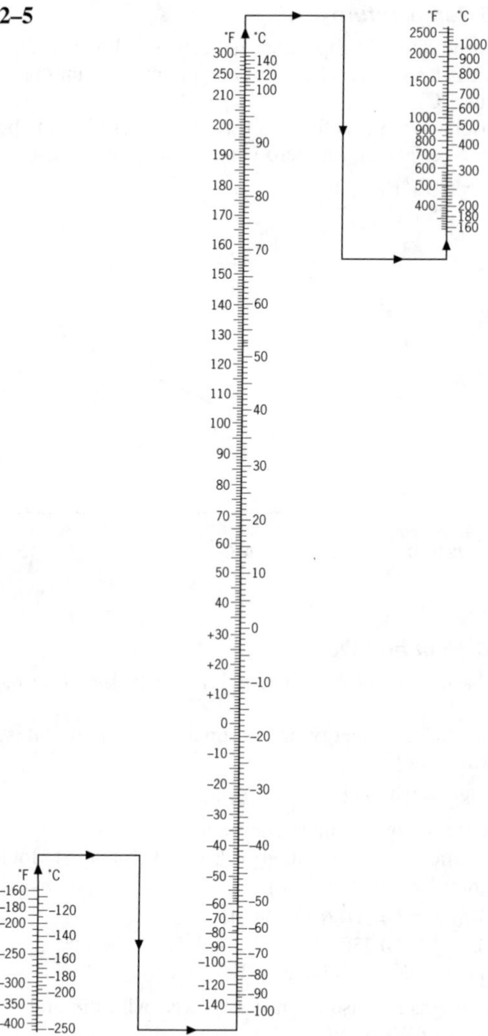

°F
°C
300
250
210
200
190
180
170
160
150
140
130
120
110
100
90
80
70
60
50
40
+30
+20
+10
0
-10
-20
-30
-40
-50
-60
-70
-80
-90
-100
-120
-140
140
120
100
90
80
70
60
50
40
30
20
10
0
-10
-20
-30
-40
-50
-60
-70
-80
-90
-100
2500
2000
1500
1000
900
800
700
600
500
400
1000
900
800
700
600
500
400
300
200
180
160
-160
-180
-200
-250
-300
-350
-400
-120
-140
-160
-180
-200
-250

Table 2.6 Heat

	Btu	*ft-1b*	*hp-h*	*cal*	*J*	*kW-h*
1 British thermal unit (Btu)	1	777.9	3.929×10^{-4}	252	1055	2.93×10^{-4}
1 foot-pound (ft-lb)	1.285×10^{-3}	1	5.051×10^{-7}	0.3239	1.356	3.766×10^{-7}
1 horsepower-hour (hp-h)	2545	1.98×10^{6}	1	6.414×10^{5}	2.685×10^{6}	0.7457
1 calorie (cal)	3.968×10^{-3}	3.087	1.559×10^{-6}	1	4.187	1.163×10^{-6}
1 Joule (J)	9.481×10^{-4}	0.7376	3.725×10^{-7}	0.2389	1	2.778×10^{-7}
1 kilowatt hour (kW-h)	3413	2.655×10^{6}	1.341	8.601×10^{5}	3.6×10^{6}	1

2.3.7 Power

Btu/h or horsepower (hp) are normally used, or, in SI, kilowatts (kW). See Table 2.7.and Fig. 2.6.

Fig. 2–6

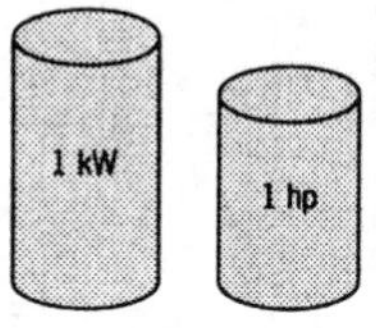

2.3.8 Flow

The basic unit of volume flowrate is US gall/min (in SI it is liters/s).

1 US gallon = 4 quarts = 128 US fluid ounces = 231 in^3.
1 US gallon = 1.2 British Imperial Gallon = 3.78833 liter (see Fig. 2.7).
1 US gallon/minute = 6.31401×10^{-5} m^3/s = 0.2273 m^3/h.
1 m^3/s = 1000 liters/s.
1 liter/s = 2.12 ft^3/min.

Fig. 2–7

2.3.9 Torque

The basic unit of torque is the foot pound (ft lbf) [in SI it is the Newton metre (Nm)]. You may also see this referred to as "moment of force" (see Fig. 2.8).

1 ft.lbf = 1.357 Nm
1 kgf.m = 9.81 Nm

Table 2.7 Power (P)

	Btu/h	*Btu/s*	*ft-lb/s*	*hp*	*cal/s*	*kW*	*W*
1 Btu/h	1	2.778×10^{-4}	0.2161	3.929×10^{-4}	7.000×10^{-2}	2.930×10^{-4}	0.2930
1 Btu/s	3600	1	777.9	1.414	252.0	1.055	1.055×10^{-3}
1ft-lb/s	4.628	1.286×10^{-3}	1	1.818×10^{-3}	0.3239	1.356×10^{-3}	1.356
1 hp	2545	0.7069	550	1	178.2	0.7457	745.7
1 cal/s	14.29	0.3950	3.087	5.613×10^{-3}	1	4.186×10^{-3}	4.186
1 kW	3413	0.9481	737.6	1.341	238.9	1	1000
1 W	3.413	9.481×10^{-4}	0.7376	1.341×10^{-3}	0.2389	0.001	1

Fig. 2–8

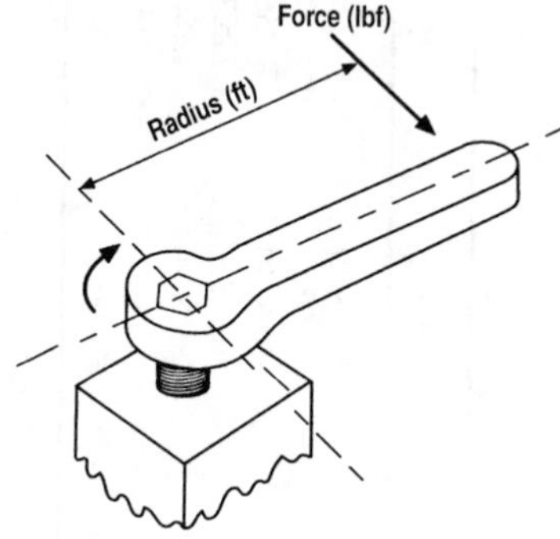

2.3.10 Stress

Stress is measured in lb/in^2—the same unit used for pressure (see Fig. 2.9), although it is a different physical quantity. In SI the basic unit is the Pascal (Pa). 1 Pa is an impractical small unit so MPa is normally used.

1 lb/in^2 = 6895 Pa
1 MPa = 1 MN/m^2 = 1 N/mm^2
1 kgf/mm^2 = 9.80665 MPa

Fig. 2–9

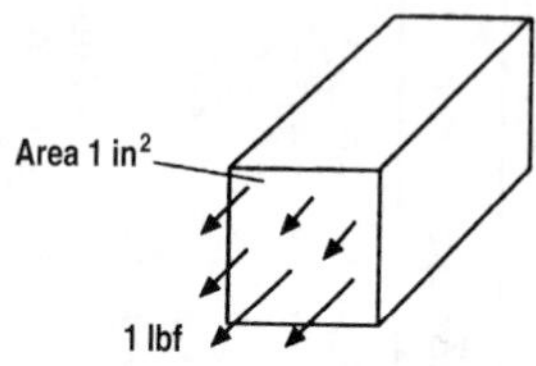

2.3.11 Linear Velocity (Speed)

The basic unit is feet per second (in SI it is m/s). See Table 2.8.

Table 2.8 Velocity (v)

Item	*ft/s*	*km/h*	*m/s*	*mile/h*	*cm/s*
1 ft per s	1	1.097	0.3048	0.6818	30.48
1 km per h	0.9113	1	0.2778	0.6214	27.78
1 m per s	3.281	3.600	1	2.237	100
1 mile per h	1.467	1.609	0.4470	1	44.70
1 cm per s	3.281×10^{-2}	3.600×10^{-2}	0.0100	2.237×10^{-2}	1

2.3.12 Acceleration

The basic unit of acceleration is feet per second squared (ft/s^2). In SI it is m/s^2.

$1\ ft/s^2 = 0.3048\ m/s^2$
$1\ m/s^2 = 3.28084\ ft/s^2$

Standard gravity (g) is normally taken as 32.1740 ft/s^2 ($9.80665 m/s^2$).

2.3.13 Angular Velocity

The basic unit is radians per second (rad/s).

1 rad/s = 0.159155 rev/s = 57.2958 degree/s

The radian is also the SI unit used for plane angles (see Fig. 2.10)

A complete circle is 2π radians
A quarter-circle (90°) is $\pi/2$ or 1.57 radians
1 degree = $\pi/180$ radians

Fig. 2–10

2.3.14 Area

The basic unit is square feet (ft^2) or square inches (in^2 or sq in). In SI it is m^2. See Table 2.9.

Other metric units of area:

Japan:

1 tsubo = 3.306 m^2
1 se = 0.9917a
1 ho-ri = 15.42 km^2

Russia:

1 kwadr. archin = 0.5058 m^2
1 kwadr. saschen = 4.5522 m^2
1 dessjatine = 1.0925 ha
1 kwadr. werst = 1.138 km^2

Small dimensions are measured in "micromeasurements" (see Fig. 2.11)

MAKING SENSE OF MICRO-INCHES (μ in)

The micro-inch (μ in) is the commonly used unit for small measures of distance:

1 microinch = 10^{-6} inches = 25.4 micrometers ("*microns*")

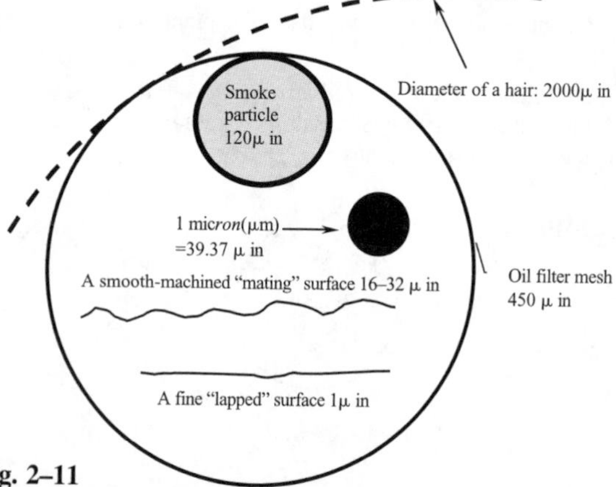

Fig. 2–11

Table 2.9 Area (A)

Unit	*sq in*	*sq ft*	*sq yd*	*sq mile*	*cm²*	*dm²*	*m²*	*a*	*ha*	*km²*
1 square inch	1	—	—	—	6.452	0.06452	—	—	—	—
1 square foot	144	1	0.1111	—	929	9.29	0.0929	—	—	—
1 square yard	1296	9	1	—	8361	83.61	0.8361	—	—	—
1 square mile	—	—	—	1	—	—	—	—	259	2.59
1 cm^2	0.155	—	—	—	1	0.01	—	—	—	—
1 dm^2	15.5	0.1076	0.01196	—	100	1	0.01	—	—	—
1 m^2	1550	10.76	1.196	—	10000	100	1	0.01	—	—
1 are (a)	—	1076	119.6	—	—	10000	100	1	0.01	—
1 hectare (ha)a	—	—	—	—	—	—	10000	100	1	0.01
1 km^2	—	—	—	0.3861	—	—	—	10000	100	1

2.3.15 Viscosity

Dynamic viscosity (μ) is measured in lbf.s/ft^2 or in the SI system in Ns/m^2 or Pascal seconds (Pa s).

1 lbf.s/ft^2 = 4.882 kgf.s/m^2 = 4.882 Pa s.
1 Pa s = 1 N s/m^2 = 1 kg/m s

A common unit from the cgs units system is the centipoise (cP). See Table 2.10.

Table 2.10 Dynamic Viscosity (μ)

Unit	*lbf-s/ft^2*	*centipoise*	*poise*	*kgf-s/m^2*
1 lb (force)-s per ft^2	1	4.788×10^4	4.788×10^2	4.882
1 centipoise	2.089×10^{-5}	1	10^{-2}	1.020×10^{-4}
1 poise	2.089×10^{-3}	100	1	1.020×10^{-2}
1 N-s per m^2	0.2048	9.807×10^3	98.07	1

Kinematic viscosity (ν) is a function of dynamic viscosity.

Kinematic viscosity = dynamic viscosity/density, i.e., $\nu = \mu/\rho$

The basic unit is ft^2/s. Other units such as Saybolt Seconds Universal (SSU) are also used.

1 m^2/s = 10.7639 ft^2/s = 5.58001×10^6 in^2/h
1 Stoke (St) = 100 centistokes (cSt) = 10^{-4} m^2/s
1 St $\cong$ 0.00226 (SSU) − 1.95/(SSU) for 32 < SSU < 100 seconds.
1 St $\cong$ 0.00220 (SSU) − 1.35/(SSU) for SSU < 100 seconds.

2.4 Consistency of Units

Within any system of units, the consistency of units forms a "quick check" of the validity of equations. The units must match on both sides.

Example:

To check kinematic viscosity (ν)

$$= \frac{\text{dynamic viscosity } (\mu)}{\text{density } (\rho)} = \mu \times 1/\rho$$

$$=> \frac{\text{ft}^2}{\text{s}} = \frac{\text{lbf.s}}{\text{ft}^2} \times \frac{\text{ft}^4}{\text{lbf.s}^2}$$

Canceling gives $\frac{ft^2}{s} = \frac{s.\,ft^4}{s^2.ft^2} = \frac{ft^2}{s}$

OK, units match.

2.5 Foolproof Conversions: Using Unity Brackets

When converting between units it is easy to make mistakes by dividing by a conversion factor instead of multiplying, or vice versa. The best way to avoid this is by using the technique of unity brackets.

A unity bracket is a term consisting of a numerator and denominator in different units which has a value of unity.

e.g. $\left[\frac{2.205\text{ lb}}{\text{kg}}\right]$ or $\left[\frac{\text{kg}}{2.205\text{ lb}}\right]$ are unity brackets

as are $\left[\frac{25.4\text{ mm}}{\text{in}}\right]$ or $\left[\frac{\text{in}}{25.4\text{ mm}}\right]$ or $\left[\frac{\text{Atmosphere}}{101325\text{ Pa}}\right]$

Remember that as the value of the bracket is unity it has no effect on any term that multiplies.

Example: Convert the density of steel $\rho = 0.29$ lb/in^3 to kg/m^3

Step 1: State the initial value:

$$\rho = \frac{0.29\text{ lb}}{\text{in}^3}$$

Step 2: Apply the "weight" unity bracket:

$$\rho = \frac{0.29\text{ lb}}{\text{in}^3}\left[\frac{\text{kg}}{2.205\text{ lb}}\right]$$

Step 3: Then apply the "dimension" unity brackets (cubed):

$$\rho = \frac{0.29\text{ lb}}{\text{in}^3}\left[\frac{\text{kg}}{2.205\text{ lb}}\right]\left[\frac{\text{in}}{25.4\text{ mm}}\right]^3\left[\frac{1000\text{ mm}}{\text{m}}\right]^3$$

Step 4: Expand and cancel:

$$\rho = \frac{0.29\text{ lb}}{\text{in}^3}\left[\frac{\text{kg}}{2.205\text{ lb}}\right]\left[\frac{\text{in}^3}{(25.4)^3\text{ mm}^3}\right]\left[\frac{(1000)^3\text{ mm}^3}{\text{m}^3}\right]$$

$$\rho = \frac{0.29kg\ (1000)^3}{2.205(25.4)^3m^3}$$

$\rho = 8025.8$ kg/m^3: Answer

* Take care to use the correct algebraic rules for the expansion: e.g.,

$(\text{a.b})^N = \text{a}^N.\text{b}^N$ *not* a.b^N

So, e.g.,

$$\left[\frac{1000\ \text{mm}}{\text{m}}\right]^3 \text{ expands to } \frac{(1000)^3.(\text{mm})^3}{(\text{m})^3}$$

Unity brackets can be used for all units conversions provided you follow the rules for algebra correctly.

2.6 Imperial—Metric Conversions

See Table 2.11

Table 2.11 Imperial—Metric Conversions

Fraction (in)	*Decimal (in)*	*Millimeter (mm)*
1/64	0.01562	0.39687
1/32	0.03125	0.79375
3/64	0.04687	1.19062
1/16	0.06250	1.58750
5/64	0.07812	1.98437
3/32	0.09375	2.38125
7/64	0.10937	2.77812
1/8	0.12500	3.17500
9/64	0.14062	3.57187
5/32	0.15625	3.96875
11/64	0.17187	4.36562
3/16	0.18750	4.76250
13/64	0.20312	5.15937
7/32	0.21875	5.55625
15/64	0.23437	5.95312
1/4	0.25000	6.35000
17/64	0.26562	6.74687
9/32	0.28125	7.14375
19/64	0.29687	5.54062
15/16	0.31250	7.93750
21/64	0.32812	8.33437
11/32	0.34375	8.73125
23/64	0.35937	9.12812
3/8	0.37500	9.52500
25/64	0.39062	9.92187
13/32	0.40625	10.31875
27/64	0.42187	10.71562
7/16	0.43750	11.11250

Table 2.11 Imperial—Metric Conversions (cont.)

Fraction (in)	*Decimal (in)*	*Millimeter (mm)*
29/64	0.45312	11.50937
15/32	0.46875	11.90625
31/64	0.48437	12.30312
1/2	0.50000	12.70000
33/64	0.51562	13.09687
17/32	0.53125	13.49375
35/64	0.54687	13.89062
9/16	0.56250	14.28750
37/64	0.57812	14.68437
19/32	0.59375	15.08125
39/64	0.60937	15.47812
5/8	0.62500	15.87500
41/64	0.64062	16.27187
21/32	0.65625	16.66875
43/64	0.67187	17.06562
11/16	0.68750	17.46250
45/64	0.70312	17.85937
23/32	0.71875	18.25625
47/64	0.73437	18.65312
3/4	0.75000	19.05000
49/64	0.76562	19.44687
25/32	0.78125	19.84375
51/64	0.79687	20.24062
13/16	0.81250	20.63750
53/64	0.82812	21.03437
27/32	0.84375	21.43125
55/64	0.85937	21.82812
7/8	0.87500	22.22500
57/64	0.89062	22.62187
29/32	0.90625	23.01875
59/64	0.92187	23.41562
15/16	0.93750	23.81250
61/64	0.95312	24.20937
31/12	0.96875	24.60625
63/64	0.98437	25.00312
1	1.00000	25.40000

2.7 Dimensional Analysis

2.7.1 Dimensional Analysis (DA)—What Is It?

DA is a technique based on the idea that one physical quantity is related to others in a precise mathematical way.
It is used for:

- Checking the validity of equations
- Finding the arrangement of variables in a formula
- Helping to tackle problems that do not possess a compete theoretical solution—particularly those involving fluid mechanics.

2.7.2 Primary and Secondary Quantities

These are quantities that are absolutely independent of one another. They are:

M Mass
L Length
T Time

For example: Velocity (v) is represented by length divided by time, and this is shown by:

$[v] = \frac{L}{T}$ note the square brackets denoting "the dimensions of."

Quantity	***Dimensions***
Mass (m)	M
Length (l)	L
Time (t)	T
Area (a)	L^2
Volume (v)	L^3
First moment of area	L^3
Second moment of area	L^4
Velocity (v)	LT^{-1}
Acceleration (a)	LT^{-2}
Angular velocity	T^{-1}
Angular acceleration	T^{-2}
Frequency (f)	T^{-1}
Force (F)	MLT^{-2}
Stress (Pressure), S (P)	$ML^{-1}T^{-2}$

Torque (T)	ML^2T^{-2}
Modulus of elasticity (E)	$ML^{-1}T^{-2}$
Work (W)	ML^2T^{-2}
Power (P)	ML^2T^{-3}
Density (ρ)	ML^{-3}
Dynamic viscosity (μ)	$ML^{-1}T^{-1}$
Kinematic viscosity (ν)	L^2T^{-1}

Hence velocity is called a secondary quantity because it can be expressed in terms of primary quantities.

An example of deriving formulae using DA

To find the formulae for periodic time (t) of a simple pendulum we can assume that t is related in some way to its mass (m), length of string (l) and acceleration due to gravity (g).

i.e., $t = \phi\{m,l,g\}$

Introducing a numerical constant C and some possible exponentials gives:

$$t = Cm^a l^b g^d$$

C is a dimensionless constant, so, in Dimensional Analysis terms, this equation becomes

$$[t] = [m^a l^b g^d]$$

Substitute primary dimensions give:

$$T = M^a L^b (LT^{-2})^d$$

$$= M^a L^{b+d} T^{-2d}$$

In order for the equation to balance

For M, a must = 0
For L, $b + d = 0$
For T, $-2d = 1$

Giving $b = \frac{1}{2}$ and $d = -\frac{1}{2}$

So we know the formula is now written:

$$t = C\, l^{1/2}\, g^{-1/2}$$

or

$$t = C\sqrt{\frac{l}{g}}\text{: The Answer}$$

Note how dimensional analysis can give the "form" of the formula but not the numerical value of the constant C.

Note also how the technique has shown that the mass (m) of the pendulum bob does not affect the periodic time (t) (i.e., because a = 0).

2.8 Useful References

For links to "The Reference Desk," a web site containing over 6000 on-line units conversions "calculators," go to: www.flinthills.com/~ramsdale/EngZone/refer.htm.

For the United States Metric Association, go to: http://lamar.colostate.edu/~hillger/.

This site contains links to over 20 units-related sites.

For guidance on correct units usage go to: http://lamar.colostate.edu/~hillger/correct.htm.

Standards: Units

1. ASTM/IEEE SI 10: 1997: *Use of the SI System of Units* (replaces ASTM E380 and IEEE 268).
2. Taylor B. N. *Guide for the Use of the International System of Units* (SI): 1995. NIST special publication N° 8111.
3. Federal Standard 376B: 1993: *Preferred Metric Units for General Use By the Federal Government.* General Services Administrations, Washington, DC 20406.

Section 3

Engineering Design—Process and Principles

3.1 Engineering Problem-Solving

Engineering is all about solving problems. Engineering design, in particular, is a complex series of events that can involve logic, uncertainty and paradox, often at the same time. There are a few "common-denominator" observations that can be made about problems in general (see Table 3.1):

Table 3.1 Engineering Problems

ENGINEERING PROBLEMS ARE:
• **Multidisciplinary** Discipline definitions are largely artificial; there are no discrete boundaries, as such, in the physical world.
• **Nested** Every part of an engineering problem contains, and is contained within, other problems. This is the property of *interrelatedness*.
• **Iterative** The final solution rarely arrives at once. The solution process is a loop.
• **Full of complexity** So you can't expect them to be simple.

3.2 Problem Types and Methodologies

Engineering problems can be divided into three main types, each with its own characteristics and methodology for finding the best solution. A methodology is a structured way of doing things. It reduces the complexity of a problem to a level you can handle.

3.2.1 Type 1: Linear Technical Problems

These consist of a basic chain of quantitative technical steps (Fig. 3.1), mainly calculations, supported by robust engineering and physical laws. There is substantial "given" information in a form that can be readily used. Note how the problem-solving

Fig. 3–1

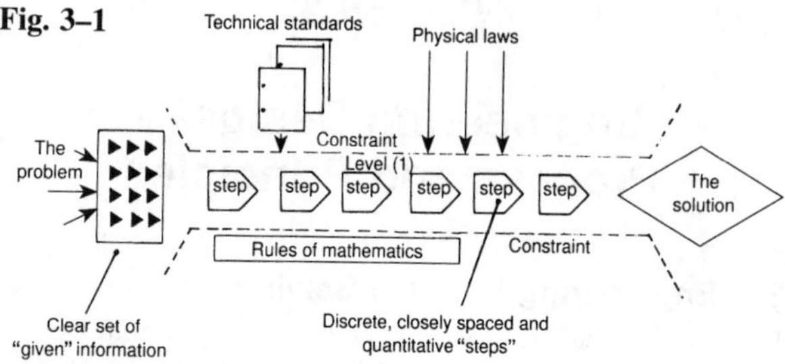

process is *linear*—each quantitative step follows on from the last and there are few, if any, iterative or retrospective activities. The solution methodology involves rigorous and accurate use of calculations and theory. Rough approximations and "order of magnitude" estimates are not good enough.

3.2.2 Type 2: Linear Procedural Problems

Their main feature is the existence of procedural constraints controlling what can be done to further define the problem and then solve it. Don't confuse these with administrative constraints, which are established procedural constraints of the *technical* world (Fig 3.2). The methodology is to use procedural techniques to solve the problem rather than approach it in a purely technical way. The problem is still in linear form, so you have to work through the steps one by one, without being retrospective (or you will lose confidence).

Fig. 3–2

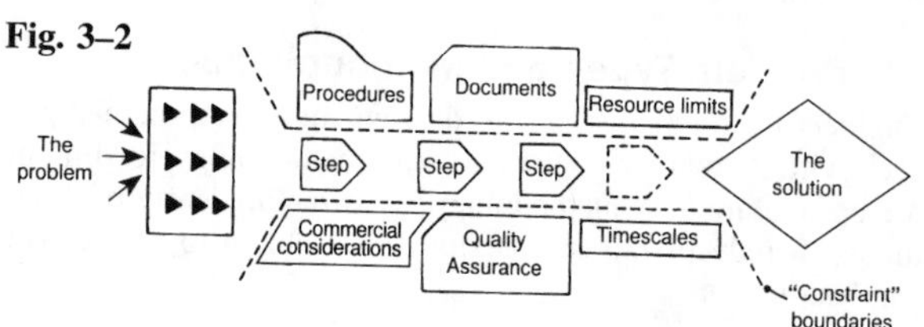

3.2.3 Type 3: Closed Problems

These look short and simple but are crammed with hidden complexity. Inside, they consist of a system of both technical mini-problems and awkward procedural constraints (Fig. 3.3)

Fig. 3–3

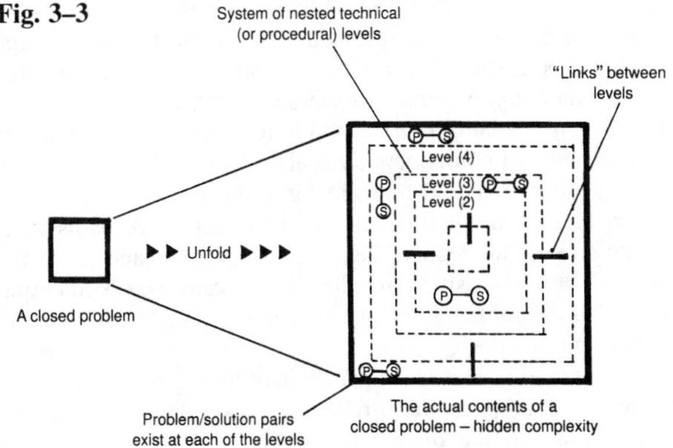

The methodology involves "opening up" the problem to reveal its complexity before you can solve it. Some hints are:

- Look for *common* nesting levels—you can anticipate these with practice.
- List the *variables and technical parameters* that you feel might be involved—then think for yourself in a pro-active way.
- Think *around* the problem, looking hard for the complexity (you will be revealing it, not introducing it, because it is there already).
- Use *group input*—closed problems do not respond well to an individual approach. A group of minds can form a richer picture of a problem than can one.

Remember the rule: Decide what type of problem you are looking at before you try to solve it.

3.3 Design Principles

Engineering design is a complex activity. It is often iterative, involving going back on old ideas until the best solution presents itself. There are, however, five well-proven principles of functional design that should be considered during the design process of any engineering product.

Clarity of function. This means that every function in a design should be achieved in a clear and simple way, i.e., without redundant components or excessive complexity.

The principle of uniformity. Good functional design encourages uniformity of component sizes and sections. Any variety that is introduced should be there for a *reason.*

Short force paths. It is always best to keep force paths short and direct. This reduces bending fiber stresses and saves material. "Local closure" (in which forces cancel each other out) is also desirable—it reduces the number of "wasted" components in a design.

Least constraint. This is the principle of letting components "go free" if at all possible. It reduces stresses due to thermal expansions and unavoidable distortions.

Use elastic design. Good elastic design avoids "competition" between rigid components, which can cause distortion and stresses. The idea is to allow components to distort in a natural way, if that is their function.

3.4 The Engineering Design Process

The *process* of engineering design is a complex and interrelated set of activities. Much has been written about how the design process works both in theory and in practice.

There is general consensus that:

Design is: the use of:

- Scientific principles
  \+
- Technical information
  \+
- Imagination

Designs are hardly ever permanent. All products around us change—sometimes gradually and sometimes in major noticeable steps—so the design process is also *continuous.* Within these points of general agreement there are various types of thought on how the process works.

3.5 Design as a Systematic Activity

This is a well-developed concept, which conceives the process as a basically linear series of steps contained within a total context or framework (see Fig. 3.4).

Fig. 3–4

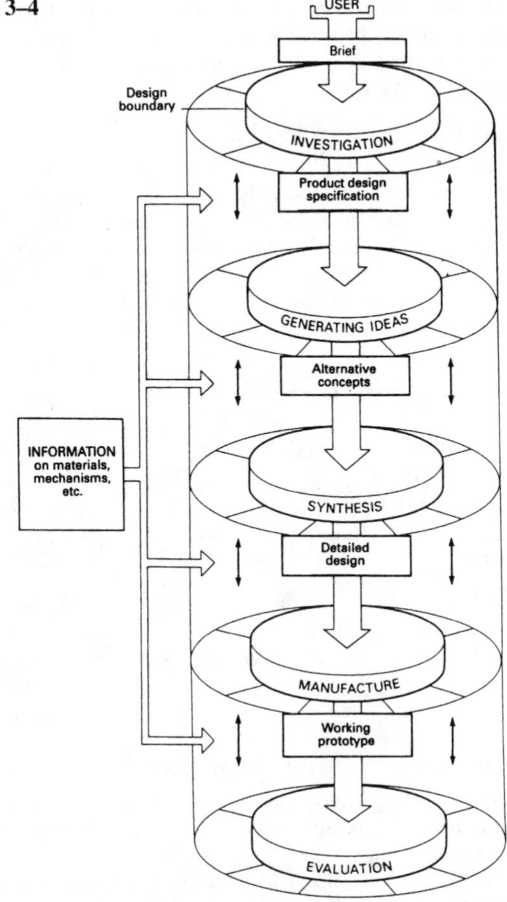

Design systemic activity model (overall concept adapted from the model used by SEED in their Curriculum for Design publications)

A central design core consists of the key stages of investigation, generating ideas, synthesis, manufacture, and evaluation. The synthesis stage is important—this is where all the technical facets of the design are brought together and formed into a final product design specification (known as the PDS). The design core is enclosed within a boundary, containing all the other factors and constraints that need to be considered. This is a disciplined and structured approach to the design process. It sees everything as a series of logical steps situated between a beginning and an end.

3.6 The Innovation Model

In contrast, this approach sees the design process as being circular or cyclic rather than strictly sequential. The process (consisting of basically the same five steps as the "systematic" approach) goes round and round, continually refining existing ideas and generating new ones (see Fig. 3.5). The activity is, however, innovation-based—it is *creativity* rather than rigor that is the key to the process.

Fig. 3–5

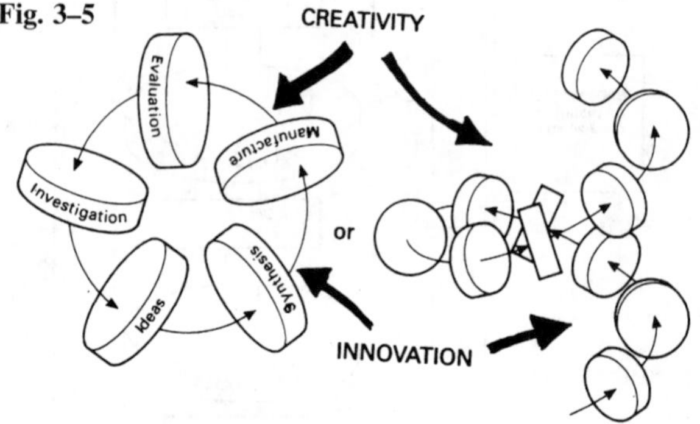

Important elements of the creative process are:

—*Lateral thinking* Conventional judgment is "put on hold" while creative processes such as brainstorming help generate new ideas.

—*Using chance* This means using a liberal approach—allowing chance to play its part (X-rays and penicillin were both discovered like this).

—*Analogy* Using analogies can help creativity, particularly in complex technical subjects.

Both approaches contain valid points. They both rely heavily on the availability of good technical information and both are *thorough* processes—looking carefully at the engineering detail of the design produced. Creativity does not have to infer a half-baked idea, or shoddiness.

3.7 The Product Design Specification (PDS)

Whatever form the design process takes, it ends with a PDS (see Fig. 3.6). This sets out broad design parameters for the designed product and sits one step "above" the detailed engineering specification (see Table 3.2).

Table 3.2 The Product Design Specification (PDS) Checklist

• Quantity	• Packing and shipping
• Product life span	• Quality
• Materials	• Reliability
• Ergonomics	• Patents
• Standardization	• Safety
• Aesthetics/finish	• Test requirements
• Service life	• Color
• Performance	• Assembly
• Product cost	• Trade marks
• Production timescale	• Value analysis
• Customer preferences	• Competing products
• Manufacture process	• Environmental factors
• Size	• Corrosion
• Disposal	• Noise levels
• Market constraints	• Documentation
• Weight	• Balance and inertia
• Maintenance	• Storage

3.8 Creativity Tools

Creativity is important in all facets of engineering design. Many of the developments in creative thinking, however, come from areas outside the engineering field. Fig. 3.7 shows five creativity tools.

Fig. 3–6

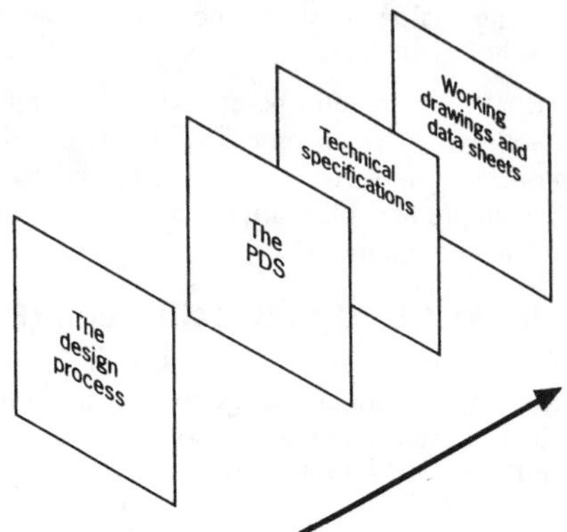

Fig. 3–7

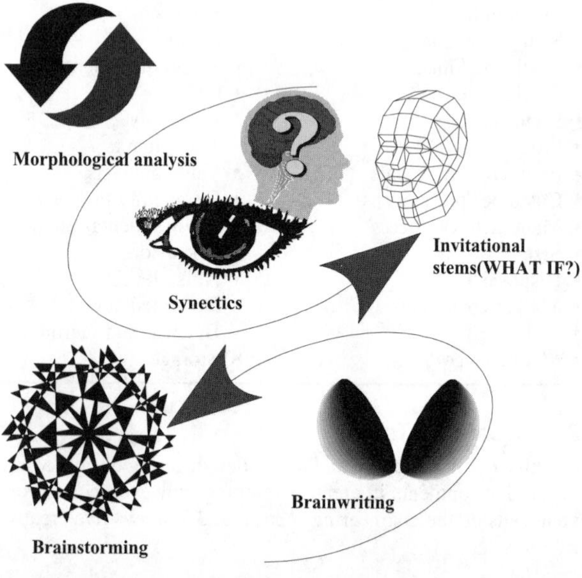

Morphological analysis

Synectics

Invitational stems(WHAT IF?)

Brainwriting

Brainstorming

- **Brainstorming** Ideas are put forward by a group of people in a "freewheeling" manner. Judgment of all ideas is deferred and absolutely no criticism is allowed. This helps stimulate originality.
- **Brainwriting** A version of brainstorming in which people contribute their ideas anonymously on slips of paper or a worksheet. These are then exchanged and people then develop (again anonymously) each other's ideas in novel ways.
- **Synectics™** A specialized technique involving the joining together of existing, apparently unrelated, ideas to reveal new perspectives or solutions to a design problem.
- **Morphological analysis** This is a formal, structured method of solving design problems using matrix analysis
- **Invitational stems (wishful thinking)** A loose and open creative process encouraged by asking questions such as "wouldn't it be nice if" or "what if material cost wasn't a problem here?"

3.9 Useful References

1. For a good four-page summary of various design methodologies and procedural models, http://itri.loyola.edu/polymers/c/_S3.htm.
2. For a good six-page paper on engineering design processes, problem-solving, and creativity: http://fre.www.ecn.purdue./edu/v1/asee/fre95/3a5/3a54/3a54.htm .
3. For a bibliography of conceptual design: http://akao/arc.nasa.gov/dfc/biblio/dfcdBiblio.html.
4. For a list of publications concerning creative design methodologies: http://www.isd.uni.stuttgart.de/~rudolph/engdesign_publications.html.

Other Useful References

de Bono. E: *Serious Creativity—Using the Power of Lateral Thinking to Create New Ideas:* 1992. Pub: Harper Collins Inc. NY.

Section 4

Basic Mechanical Design

4.1 Engineering Abbreviations

The following abbreviations, based on the published standard *ANSI/ASME Y14.5 1994: Dimensioning and Tolerancing*, are in common use in engineering drawings and specifications (Table 4.1).

Table 4.1 Engineering Abbreviations

Abbreviation	*Meaning*
ANSI	American National Standards Institute
ASA	American Standards Association
ASME	American Society of Mechanical Engineers
AVG	average
CBORE	counterbore
CDRILL	counterdrill
CL	centerline
CSK	countersink
FIM	full indicator movement
FIR	full indicator reading
GD&T	geometric dimensioning and tolerancing
ISO	International Standards Organization
LMC	least material condition
MAX	maximum
MDD	master dimension definition
MDS	master dimension surface
MIN	minimum
mm	millimeter
MMC	maximum material condition
PORM	plus or minus
R	radius
REF	reference
REQD	required
RFS	regardless of feature size
SEP REQT	separate requirement

Table 4.1 Engineering Abbreviations (cont.)

Abbreviation	*Meaning*
SI	Système International (the metric system)
SR	spherical radius
SURF	surface
THRU	through
TIR	total indicator reading
TOL	tolerance

4.2 Preferred Numbers and Preferred Sizes

Preferred numbers are derived from geometric series, in which each term is a uniform percentage larger than its predecessor. The first five principal series (named the "R" series) are shown in Fig 4.1.

Fig. 4–1

Series	*Basis*	*Ratio of terms (% increase)*
R5	$\sqrt[5]{10}$	1.58(58%)
R10	$\sqrt[10]{10}$	1.26(26%)
R20	$\sqrt[20]{10}$	1.12(12%)
R40	$\sqrt[40]{10}$	1.06(6%)
R80	$\sqrt[80]{10}$	1.03(3%)

Preferred numbers are taken as the basis for ranges of linear sizes of components, often being rounded up or down for convenience. Figure 4.2 shows the development of the R5 and R10 series.

Fig. 4–2

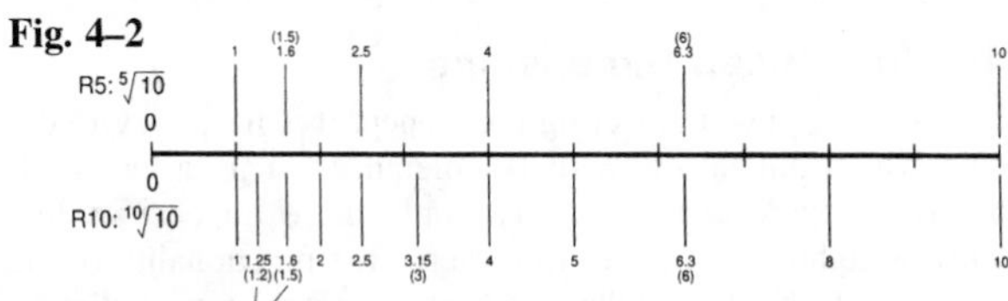

4.3 Datums and Tolerances—Principles

A *datum* is a reference point or surface from which all other dimensions of a component are taken; these other dimensions are said to be *referred to* the datum. In most practical designs, a datum surface is normally used, this generally being one of the surfaces of the machine element itself rather than an "imaginary" surface. This means that the datum surface normally plays some important part in the operation of the elements—it is usually machined and may be a mating surface or a locating face between elements, or similar (see Fig. 4.3). Simple machine mechanisms do not *always* need datums; it depends on what the elements do and how complicated the mechanism assembly is.

Fig. 4–3

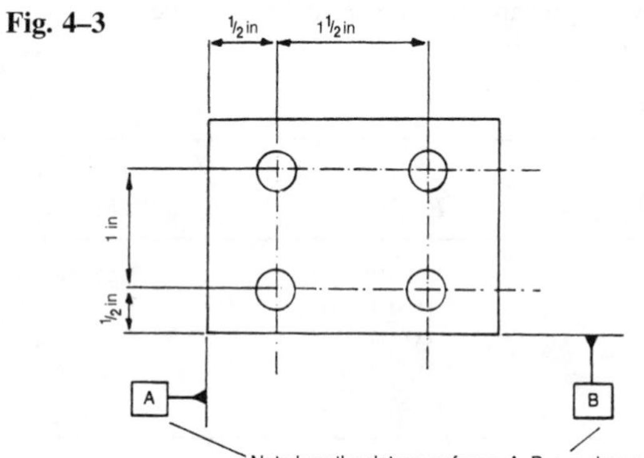

A *tolerance* is the allowable variation of a linear or angular dimension about its "perfect" value. ANSI/ASME Y14.5: 1994 contains accepted methods and symbols (see Fig. 4.4).

4.4 Toleranced Dimensions

In designing any engineering component it is necessary to decide which dimensions will be toleranced. This is predominantly an exercise in necessity—only those dimensions that *must* be tightly controlled, to preserve the functionality of the component, should be toleranced. Too many toleranced dimen-

See also:
ANSI/ASME Y14.5

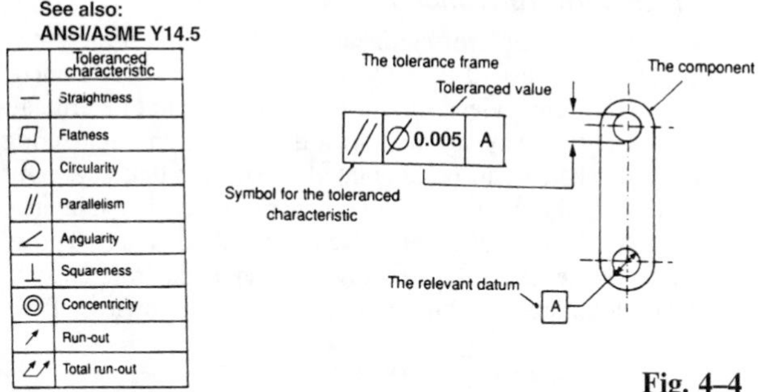

Fig. 4–4

sions will increase significantly the manufacturing costs and may result in "tolerance clash," where a dimension derived from other toleranced dimensions can have several contradictory values (see Fig. 4.5).

Fig. 4–5

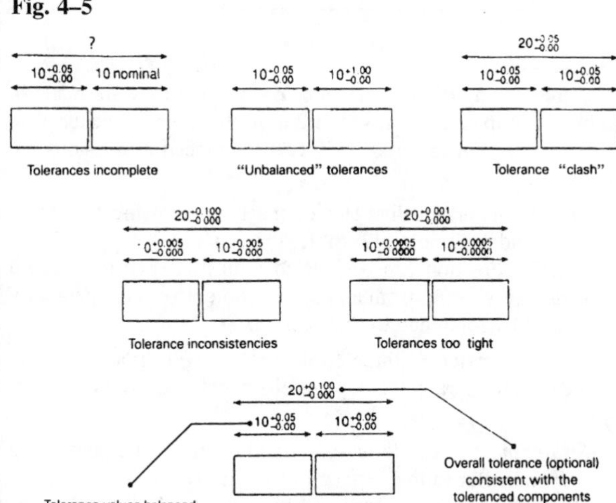

4.4.1 General Tolerances

It is a sound principle of engineering practice that in any machine design there will be only a small number of toleranced features. The remainder of the dimensions will not be critical.

There are two ways to deal with this: First, an engineering drawing or sketch can be annotated to specify that a *general tolerance* should apply to features where no specific tolerance is mentioned. This is often expressed as ± 0.020 in or "20 mils" (0.5 mm). Alternatively, the drawing can make reference to "typical" tolerances for linear dimensions (as shown in Table 4.2).

Table 4.2 Typical tolerances for linear dimensions

Dimension	*Tolerance (mils)*
1/40 in - 1/4 in (0.63 mm–6.36 mm)	± 4 (0.1 mm)
1/4 in - 1 1/2 in (6.36 mm–38.1 mm)	± 8 (0.2 mm)
1 1/2 in - 5 in (38.1 mm –127 mm)	± 12 (0.3 mm)
5 in - 12 1/2 in (1207 mm–317.5 mm)	± 20 (0.5 mm)
12 1/2 in - 40 in (317.5–1016 mm)	± 32 (0.8 mm)

4.4.2 Holes

The tolerancing of holes depends on whether they are made in thin sheet [up to about 1/8 in (3.2 mm) thick] or in thicker plate material. In thin material, only two toleranced dimensions are required:

—*Size* A toleranced diameter of the hole, showing the maximum and minimum allowable dimensions.

—*Position* Position can be located with reference to a datum and/or its spacing from an adjacent hole. Holes are generally spaced by reference to their centers.

For thicker material, three further toleranced dimensions become relevant: straightness, parallelism and squareness (see Fig. 4.6).

—*Straightness* A hole or shaft can be *straight* without being perpendicular to the surface of the material.

—*Parallelism* This is particularly relevant to holes and is important when there is a mating hole-to-shaft fit.

Fig. 4–6

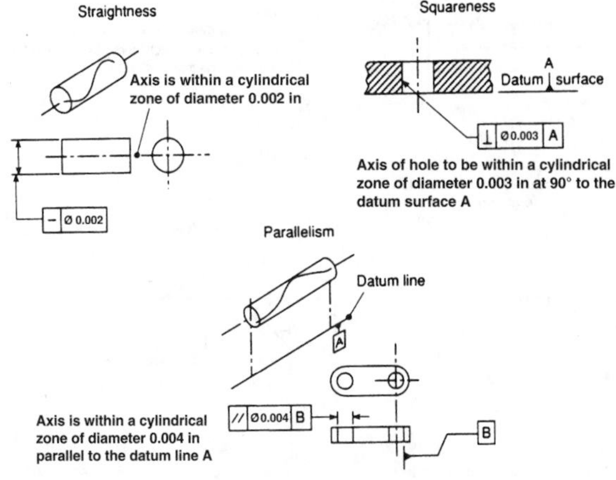

Straightness, parallelism, and squareness

—*Squareness* The formal term for this is perpendicularity. Simplistically, it refers to the squareness of the axis of a hole to the datum surface of the material through which the hole is made.

4.4.3 Screw Threads

There is a well-established system of tolerancing adopted by ANSI/ASME, International Standard Organizations and manufacturing industry. This system uses the two complementary elements of fundamental deviation and tolerance range to define fully the tolerance of a single component. It can be applied easily to components, such as screw threads, which join or mate together (see Fig. 4.7).

—*Fundamental deviation* (FD) is the distance (or "deviation") of the nearest "end" of the tolerance band from the nominal or "basic" size of a dimension.

—*Tolerance band* (or "range") is the size of the tolerance band, i.e., the difference between the maximum and minimum acceptable size of a toleranced dimension. The size of the tolerance band, and the location of the FD, governs the system of limits and fits applied to mating parts.

Fig. 4–7

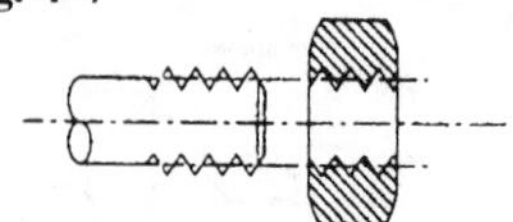

For screw threads, the tolerance layout shown applies to major, pitch, and minor diameters (although the actual values will differ)

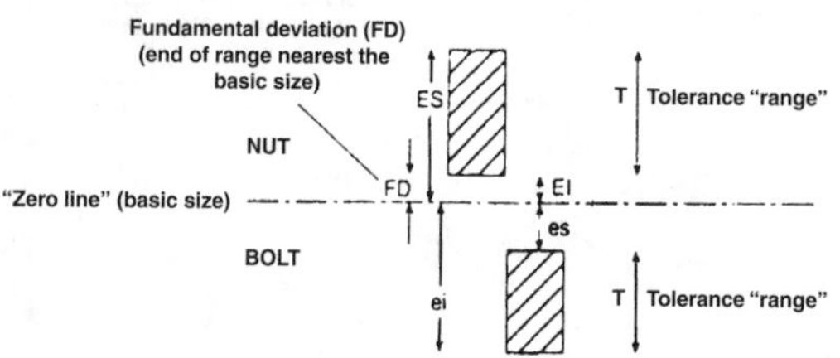

FD is designated by a letter code, e.g., g. H
Tolerance range (T) is designated by a number code, e.g., 5, 6, 7

Commonly used symbols are:
EI – lower deviation (nut)
ES – upper deviation (nut)
ei – lower deviation (bolt)
es – upper deviation (bolt)

Tolerance values have a key influence on the costs of a manufactured item so their choice must be seen in terms of economics as well as engineering practicality. Mass-produced items are competitive and price sensitive, and overtolerancing can affect the economics of a product range.

4.5 Limits and Fits

4.5.1 Principles

In machine element design there is a variety of different ways in which a shaft and hole are required to fit together. Elements such as bearings, location pins, pegs, spindles, and axles are typical examples. The shaft may be required to be a tight fit in the hole, or to be looser, giving clearance to allow easy removal or rotation. The system designed to establish a series of useful fits between shafts and holes is termed *limits and fits*. This involves a series of tolerance grades so that machine elements can be made with the correct degree of accuracy and be interchangeable with others of the same tolerance grade.

The standards ANSI B4.1/B4.3 contain the recommended tolerances for a wide range of engineering requirements. Each fit is designated by a combination of letters and numbers (see Table 4.3).

Table 4.3 Classes of Fit

1. *Loose running fit*: Class RC8 and RC9. These are used for loose "commercial-grade" components where a significant clearance is necessary.
2. *Free running fit*: Class RC7. Used for loose bearings with large temperature variations.
3. *Medium running fit*: Class RC6 and RC5. Used for bearings with high running speeds.
4. *Close running fit*: Class RC4. Used for medium-speed journal bearings.
5. *Precision running fit*: Class RC3. Used for precision and slow-speed journal bearings.
6. *Sliding fit*: Class RC2. A locational fit in which close-fitting components slide together.
7. *Close sliding fit*: Class RC1. An accurate locational fit in which close-fitting components slide together.
8. *Light drive fit*: Class FN1. A light push fit for long or slender components.
9. *Medium drive fit*: Class FN2. A light shrink-fit suitable for cast-iron components.
10. *Heavy drive fit*: Class FN3. A common shrink-fit for steel sections.
11. *Force fits*: Class FN4 and FN5. Only suitable for high-strength components.

Figure 4.8 shows the principles of a shaft/hole fit. The "zero line" indicates the basic or "nominal" size of the hole and shaft (it is the same for each) and the two shaded areas depict the tolerance zones within which the hole and shaft may vary. The hole is conventionally shown above the zero line. The algebraic difference between the basic size of a shaft or hole and its actual size is known as the *deviation*.

—It is the deviation that determines the nature of the fit between a hole and a shaft.

Fig. 4–8

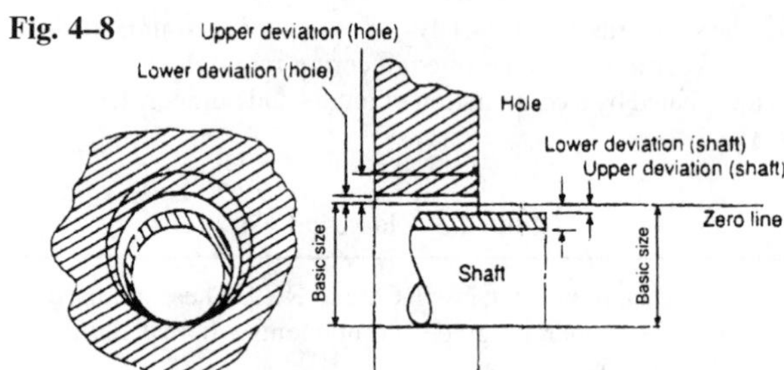

—If the deviation is small, the tolerance range will be near the basic size, giving a tight fit.
—A large deviation gives a loose fit.

Various grades of deviation are designated by letters, similar to the system of numbers used for the tolerance ranges. Shaft deviations are denoted by small letters and hole deviations by capital letters. Most general engineering uses a "hole-based" fit in which the larger part of the available tolerance is allocated to the hole (because it is more difficult to make an accurate hole) and then the shaft is made to suit, to achieve the desired fit.

Tables 4.4 and 4.5 show suggested clearance and fit dimensions for various diameters (ref: ANSI B4.1 and 4.3)

Table 4.4 Force and Shrink Fits

Nominal size range, in	*Class*				
	FN 1	*FN2*	*FN3*	*FN4*	*FN5*
0.04-0.12	0.05	0.2		0.3	0.5
	0.5	0.85		0.95	1.3
0.12-0.24	0.1	0.2		0.95	1.3
	0.6	1.0		1.2	1.7
0.24-0.40	0.1	0.4		0.6	0.5
	0.75	1.4		1.6	2.0
0.40-0.56	0.1	0.5		0.7	0.6
	0.8	1.6		1.8	2.3
0.56-0.71	0.2	0.5		0.7	0.8
	0.9	1.6		1.8	2.5

Table 4.4 Force and Shrink Fits (cont.)

Nominal size range, in	*Class*				
	FN 1	*FN2*	*FN3*	*FN4*	*FN5*
0.71-0.95	0.2	0.6		0.8	1.0
	1.1	1.9		2.1	3.0
0.95-1.19	0.3	0.6	0.8	1.0	1.3
	1.2	1.9	2.1	2.3	3.3
1.19-1.58	0.3	0.8	1.0	1.5	1.4
	1.3	2.4	2.6	3.1	4.0
1.58-1.97	0.4	0.8	1.2	1.8	2.4
	1.4	2.4	2.8	3.4	5.0
1.97-2.56	0.6	0.8	1.3	2.3	3.2
	1.8	2.7	3.2	4.2	6.2
2.56-3.15	0.7	1.0	1.8	2.8	4.2
	1.9	2.9	3.7	4.7	7.2

Limits in "mils" (0.001in)

Table 4.5 Running and Sliding Fits

Nominal size range, in	*Class*								
	RC1	*RC2*	*RC3*	*RC4*	*RC5*	*RC6*	*RC7*	*RC 8*	*RC9*
0-0.12	0.1	0.1	0.3	0.3	0.6	0.6	1.0	2.5	4.0
	0.45	0.55	0.95	1.3	1.6	2.2	2.6	5 .1	8.1
0.12-0.24	1.5	0.15	0.4	0.4	0.8	0.8	1.2	2.8	4.5
	0.5	0.65	1.2	1.6	2.0	2.7	3.1	5.8	9.0
0.24-0.40	0.2	0.2	0.5	0.5	1.0	1.0	1.6	3.0	5.0
	0.6	0.85	1.5	2.0	2.5	3.3	3.9	6.6	10.7
0.40-0.71	0.25	0.25	0.6	0.6	1.2	1.2	2.0	3.5	6.0
	0.75	0.95	1.7	2.3	2.9	3.8	4.6	7. 9	12.8
0.71-1.19	0.3	0.3	0.8	0.8	1.6	1.6	2.5	4.5	7.0
	0.95	1.2	2.1	2.8	3.6	4.8	5.7	10. 0	15.5
1.19-1.97	0.4	0.4	1.0	1.0	2.0	2.0	3.0	5.0	8.0
	1.1	1.4	2.6	3.6	4.6	6.1	7.1	11.5	18.0
1.97-3.15	0.4	0.4	1.2	1.2	2.5	2.5	4.0	6.0	9.0
	1.2	1.6	3.1	4.2	5.5	7.3	8.8	13.5	20.5
3.15-4.73	0.5	0.5	1.4	1.4	3.0	3.0	5.0	7.0	10.0
	1.5	2.0	3.7	5.0	6.6	8.7	10.7	15.5	24.0

Limits in "mils" (0.001in)

4.5.2 Metric Equivalents

The metric system (ref: ISO Standard EN 20286) *ISO "limits and fits"* uses seven popular combinations with similar definitions (see Table 4.6 and Fig. 4.9).

Table 4.6 Metric Fit Classes

1. *Easy running fit*: H11–c11, H9–d10, H9–e9. These are used for bearings where a significant clearance is necessary.
2. *Close running fit*: H8–f7, H8–g6. This only allows a small clearance, suitable for sliding spigot fits and infrequently used journal bearings. This fit is not suitable for continuously rotating bearings.
3. *Sliding fit*: H7–h6. Normally used as a locational fit in which close-fitting items slide together. It incorporates a very small clearance and can still be freely assembled and disassembled.
4. *Push fit*: H7–k6. This is a transition fit, mid-way between fits that have a guaranteed clearance and those where there is metal interference. It is used where accurate location is required, e.g., dowel and bearing inner-race fixings.
5. *Drive fit*: H7–n6. This is a tighter grade of transition fit than the H7–k6. It gives a tight assembly fit where the hole and shaft may need to be pressed together.
6. *Light press fit*: H7–p6. This is used where a hole and shaft need permanent, accurate assembly. The parts need pressing together but the fit is not so tight that it will overstress the hole bore.
7. *Press fit*: H7–s6. This is the tightest practical fit for machine elements such as bearing bushes. Larger interference fits are possible but are suitable only for large heavy engineering components.

Fig. 4–9 METRIC EQUIVALENTS

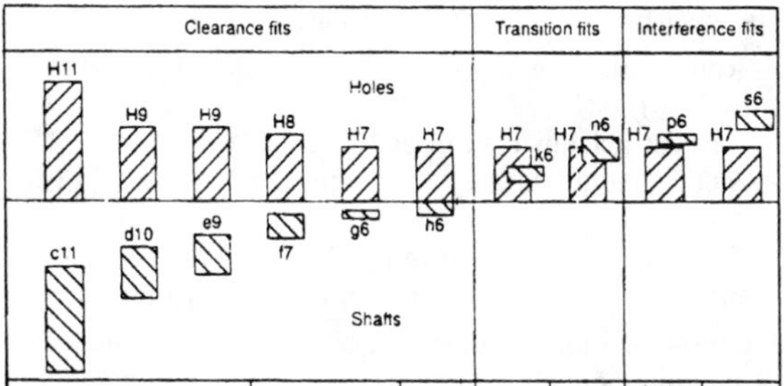

	Clearance fits												Transition fits				Interference fits			
	Easy running						Close running				Sliding		Push		Drive		Light press		Press	
Nominal size in mm	Tols*		Tols		Tols		Tols		Tols		Tols		Tols		Tols		Tols		Tols	
	H11	c11	H9	d10	H9	e9	H8	f7	H7	g6	H7	h6	H7	k6	H7	n6	H7	p6	H7	s6
6–10	+90 0	−80 −170	+36 0	−40 −98	+36 0	−25 −61	+22 0	−12 −28	+15 0	−5 −14	+15 0	−9 0	+15 0	+10 +1	+15 0	+19 +10	+15 0	+24 +15	+15 0	+32 +23
10–18	+110 0	−95 −205	+43 0	−50 −120	+43 0	−32 −75	+27 0	−16 −34	+18 0	−6 −17	+18 0	−11 0	+18 0	+12 +1	+18 0	+23 +12	+18 0	+29 +18	+18 0	+39 +28
18–30	+130 0	−110 −240	+52 0	−69 −149	+52 0	−40 −92	+33 0	−20 −41	+21 0	−7 −20	+21 0	−13 0	+21 0	+15 +2	+21 0	+28 +15	+21 0	+35 +22	+21 0	+48 +35
30–40	+140 0	−120 −280	+62	−80	+62	−50	+39	−25	+25	−9	+25	−16	+25	+18	+25	+33	+25	+42	+25	+59
40–50	+160 0	−130 −290	0	−180	0	−112	0	−50	0	−25	0	0	0	+2	0	+17	0	+26	0	+43

* Tolerance units in 0.001 mm

Ref: ISO/EN 20286

4.6 Surface Finish

Surface finish, more correctly termed "surface texture," is important for all machine elements that are produced by machining processes such as turning, grinding, shaping, or honing. This applies to surfaces that are flat or cylindrical. Surface texture is covered by its own technical standard: ASME/ANSI B46.1: 1995: *Surface Texture.* It is measured using the parameter R_a, which is a measurement of the average distance between the median line of the surface profile and its peaks and troughs, measured in micro-inches (μin). There is another system from a comparable European standard, DIN ISO 1302, which uses a system of N-numbers—it is simply a different way of describing the same thing.

4.6.1 Choice of Surface Finish: Approximations

Basic surface finish designations are:

—Rough turned, with visible tool marks: 500 μ in R_a(12.5 μm or N10)
—Smooth machined surface: 125 μ in R_a (3.2 μm or N8)
—Static mating surfaces (or datums): 63 μ in R_a (1.6 μm or N7)
—Bearing surfaces: 32 μ in R_a (0.8 μm or N6)
—Fine "lapped" surfaces: 1 μ in R_a (0.025 μm or N1)

Figure 4.10 shows comparison between the different methods of measurement.

Fig. 4–10

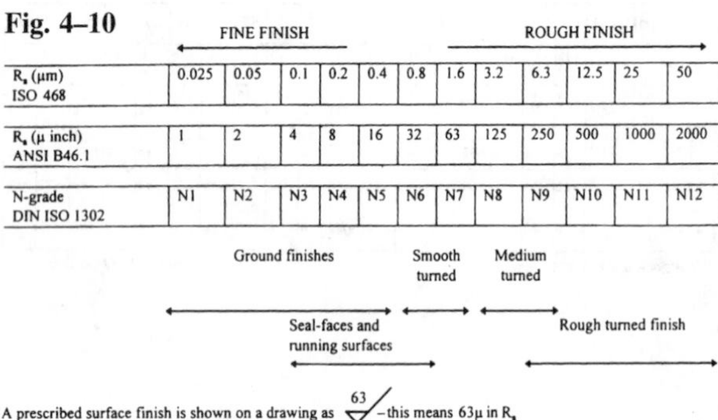

Ra (µm) ISO 468	0.025	0.05	0.1	0.2	0.4	0.8	1.6	3.2	6.3	12.5	25	50
Ra (µ inch) ANSI B46.1	1	2	4	8	16	32	63	125	250	500	1000	2000
N-grade DIN ISO 1302	N1	N2	N3	N4	N5	N6	N7	N8	N9	N10	N11	N12

Finer finishes can be produced but are more suited for precision application such as instruments. It is good practice to specify the surface finish of close-fitting surfaces of machine elements, as well as other ASME/ANSI Y 14.5 parameters such as squareness and parallelism.

4.7 Useful References

Standards: Limits, Tolerances, and Surface Texture

1. ANSI Z17.1: 1976: *Preferred Numbers.*
2. ANSI B4.2: 1999: *Preferred Metric Limits and Fits.*
3. ANSI B4.3: 1999: *General Tolerances for Metric Dimensioned Products.*

4. ANSI/ASME Y14.5.1 M: 1999: *Dimensioning and Tolerances—Mathematical Definitions of Principles.*
5. ASME B4.1: 1999: *Preferred Limits and Fits for Cylindrical Parts.*
6. ASME B46.1: 1995: *Surface Texture (Surface Roughness, Waviness, and Lay)*
7. ISO 286-1: 1988: *ISO System of Limits and Fits.*

Standards: Screw Threads

1. ASME B1.1: 1989: *Unified Inch Screw Threads (UN and UNR Forms).*
2. ASME B1.2: 1991: *Gages and Gaging for Unified Screw Threads.*
3. ASME B1.3M: 1992: *Screw Thread Gaging Systems for Dimensional Acceptability—Inch and Metric Screws.*
4. ASME B1.13: 1995: *Metric Screw Threads.*
5. ISO 5864: 1993: *ISO Inch Screw Threads—Allowances and Tolerances.*

Useful Web Sites

Geometric Dimensioning and Tolerancing Professionals (GDTP - Y14.5.2): www.asme.org/cns/departments/AccredCertif/GDTP.htm

Web information from the publication *Basics of Design Engineering*: http://machinedesign.com/bde/main.html

For wide-ranging technical information on many aspects of engineering design: http://www.flinthills.com/~ramsdale/EngZone/refer.htm

Other Useful References

Rothbart, H.A. *Mechanical Design Handbook*: 1995. Pub: McGraw-Hill.ISBN 0 07 054038 1.

Pope, E.J. *Rules of Thumb for Mechanical Engineers*: 1997. Pub: Gulf. ISBN 0 88415 790 3.

Section 5

Motion

5.1 Making Sense of Equilibrium

The concept of equilibrium lies behind many types of engineering analyses and design. Some key definition points are:

- Formally, an object is in a state of equilibrium when the forces acting on it are such that they leave it in its state of rest or uniform motion in a straight line.
- Practically, the most useful interpretation is that an object is in equilibrium when the forces acting on it are producing no tendency for the object to *move*.

Figure 5.1 shows the difference between equilibrium and non-equilibrium.

Fig. 5–1

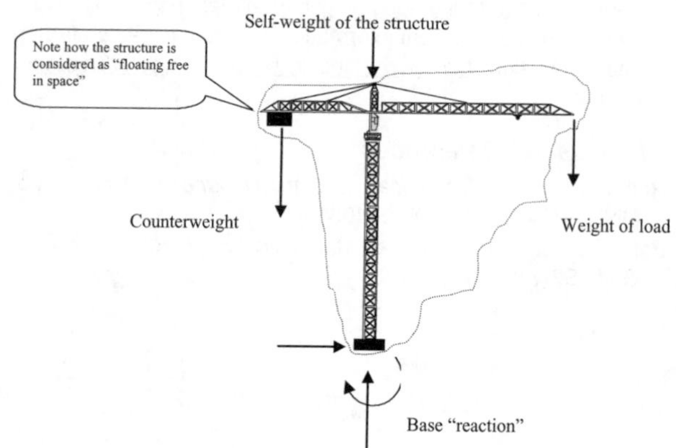

Fig. 5–1 (cont.)

NON- EQUILIBRIUM

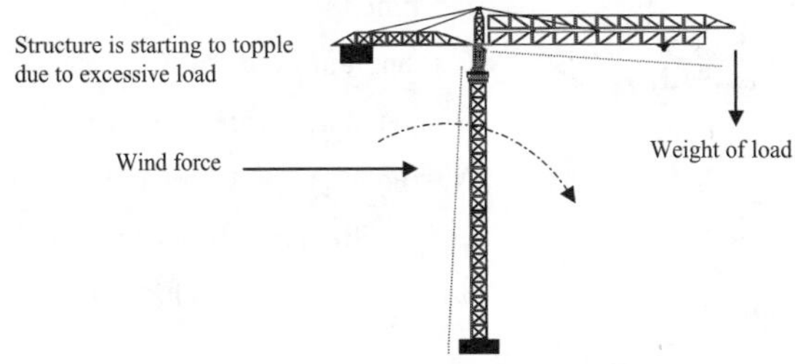

The concept of equilibrium is used to analyze engineering structures and components. By isolating a part of a structure (a joint or a member), which is in a state of equilibrium, this enables a "free body diagram" to be drawn. This helps in the analysis of the stresses (and the resulting strains) in the structure. When co-planar forces acting at a point are in equilibrium, the vector diagram closes.

5.2 Motion Equations

5.2.1 Uniformly Accelerated Motion

Bodies under uniformally accelerated motion follow the general equations:

$v = u + at$ t = time (s)

$s = ut + \frac{1}{2}at^2$ a = acceleration (ft/s^2)

$s = \frac{u + v}{2}t$ s = distance travelled (ft)

u = initial velocity (ft/s)

$v^2 = u^2 + 2as$ v = final velocity (ft/s)

5.2.2 Angular Motion

$$\omega = \frac{2\pi N}{60}$$

$$\omega_2 = \omega_1 + \alpha t$$

$$\theta = \frac{\omega_1 + \omega_2}{2} t$$

$$\omega_2{}^2 = \omega_1{}^2 + 2\alpha s$$

$$\theta = \omega_1 t + \frac{1}{2} \alpha t^2$$

t = time (s)

θ = angle moved (rad)

α = angular acceleration (rad/s^2)

N = angular speed (rev/min)

ω_1 = initial angular velocity (rad/s)

ω_2 = final angular velocity (rad/s)

5.2.3 General Motion of a Particle in a Plane

$v = \mathrm{d}s/\mathrm{d}t$

$a = \mathrm{d}v/\mathrm{d}t = \mathrm{d}_2 s/\mathrm{d}t^2$

$v = adt$

$s = vdt$

s = distance

t = time

v = velocity

a = acceleration

5.3 Newton's Laws of Motion

First law Every body will remain at rest or continue in uniform motion in a straight line until acted upon by an external force.

Second law When an external force is applied to a body of constant mass it produces an acceleration which is directly proportional to the force. i.e., Force (F) lbf = mass (m) lbm $\times$ acceleration (a) ft/s^2

Third law Every action produces an equal and opposite reaction.

Table 5.1 shows comparisons between rotational and translational motion.

Table 5.1 Comparisons: Rotational and Translational Motion

Translation		*Rotation*	
Linear displacement from a datum	x	Angular displacement	θ
Linear velocity	ν	Angular velocity	ω
Linear acceleration	$a = d\nu/dt$	Angular acceleration	$\alpha = d\omega/dt$
Kinetic energy	$KE = m\nu^2/2$	Kinetic energy	$KE = I\omega^2/2$
Momentum	$m\nu$	Momentum	$I\omega$
Newton's second law	$F = md_2x/dt^2$	Newton's second law	$M = d_2\theta/dt^2$

5.4 Simple Harmonic Motion

A particle moves with "simple harmonic motion" when it has constant angular velocity (ω) and follows a displacement pattern $x = x_0 \sin(2\pi Nt/60)$. The projected displacement, velocity, and acceleration of a point P on the x-y axes are a sinusoidal function of time (t) (see Fig. 5.2).

x_0 = Amplitude of the displacement
Angular velocity (ω) = $2\pi N/60$ = Where N is in revs/min
Periodic time (T) = $2\pi/\omega$
Velocity (v) of point A on the x axis is:
$v = ds/dt = \omega r \sin \omega t$
Acceleration (a) = $d_2s/dt^2 = dv/dt = -\omega^2 r \cos \omega t$

Fig. 5–2 Simple harmonic motion

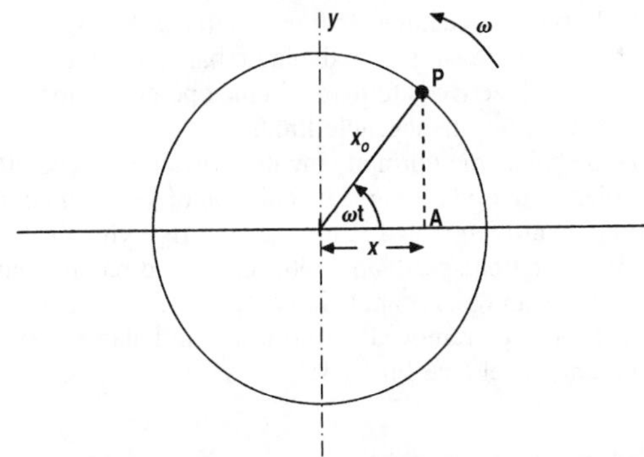

5.5 Understanding Acceleration

The dangerous thing about acceleration is that it represents a *rate of change* of speed or velocity. When this rate of change is high it puts high stresses on engineering components, causing them to deform and break. In the neat world of physical science, objects in a vacuum experience a constant acceleration (g) due to gravity of 32.1740 ft/s^2 (9.80665 m/s^2)—so if you drop a hammer and a feather they will reach the ground at the same time.

Unfortunately, you won't find many engineering products made of hammers and feathers locked inside vacuum chambers. In practice, the components of engineering machines experience acceleration many times the force of gravity so they have to be designed to resist the forces that result. Remember that these forces can be caused as a result of either linear or angular accelerations and that there is a comparison between the two as shown below :

Linear acceleration	***Angular acceleration***
$a = \dfrac{v - u}{\mathrm{t}}$ ft/s^2	$\alpha = \dfrac{\omega_2 - \omega_1}{\mathrm{t}}$ rad/s^2
so, a and α are linked by $a = \mathrm{r}\omega$	*where* a is the radius of the circle

When analyzing (or designing) any machine or mechanism think about linear accelerations *first*—they are always important.

5.6 Dynamic Balancing

Almost all rotating machines (pumps, shafts, turbines, gear sets, generators, etc.) are subject to dynamic balancing during manufacture. The objective is to maintain the operating vibration of the machine within manageable limits.

Dynamic balancing normally involves two measurement/correction planes and involves the calculation of vector quantities. The component is mounted in a balancing rig, which rotates it at near its operating speed, and both senses and records out-of-balance forces and phase angle in two planes. Balance weights are then added (or removed) to bring the imbalance forces to an acceptable level (see Fig. 5.3).

Fig. 5–3

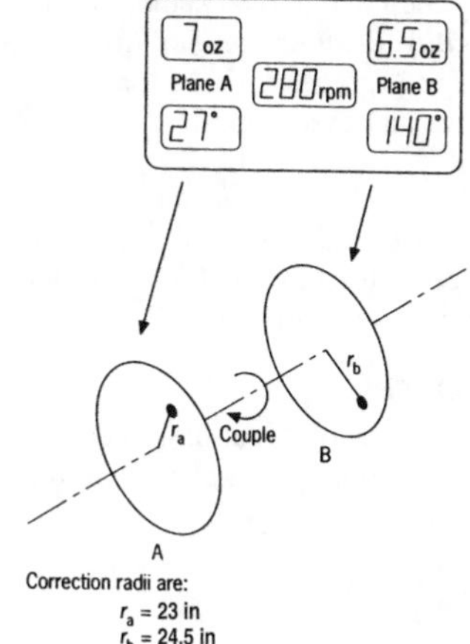

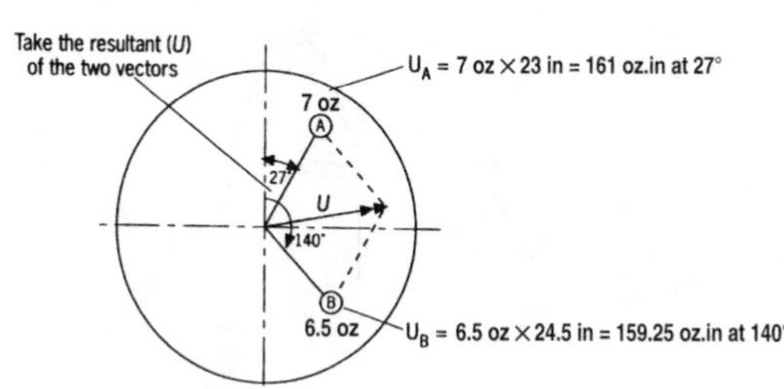

For the resultant:
U ≃ 224 oz.in at $\phi \simeq 77°$ (by graphical methods)

5.6.1 Balancing Terminology Standard ANSI S2.7: 1997

The national standard ANSI S2.7 is widely used as the basis of terminology used for balancing techniques and equipment. It covers all aspects relating to machines, rotors, balancing, and equipment.

5.6.2 Balancing Levels

ANSI/ASA standard ANSI S2.42: 1997 *Balancing of Flexible Rotors* is frequently used. This classifies rotors into groups in accordance with various balance “quality” grades. A similar approach is used by the international standard ISO 1940-1: 1984 *Balance and Quality Requirements of Rigid Rotors* and ISO 10816-1. Finer balance grades are used for precision assemblies such as instruments and gyroscopes.

5.7 Vibration

Vibration is a subset of the subject of dynamics. It has particular relevance to both structures and machinery in the way that they respond to applied disturbances.

5.7.1 General Model

The most common model of vibration is a concentrated spring-mounted mass that is subject to a disturbing force and retarding force (see Fig. 5.4).

Fig. 5–4

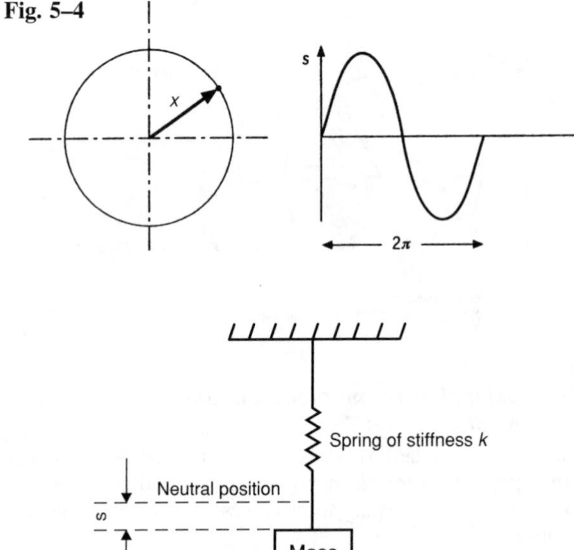

The motion is represented graphically as shown by the projection of the rotating vector x. Relevant quantities are

frequency (Hz) = $\sqrt{k/m}/2\pi$
k = spring stiffness
m = mass

The ideal case represents simple harmonic motion with the waveform being sinusoidal. Hence the motion follows the general pattern:

Vibration displacement (amplitude) = s
Vibration velocity = $v = \mathrm{d}s/\mathrm{dt}$
Vibration acceleration = $a = \mathrm{d}v/\mathrm{dt}$

5.7.2 Machine Vibration

There are two types of vibration relevant to rotating machines

—Bearing *housing* vibration. This is assumed to be sinusoidal. It normally uses the velocity (V_{rms}) parameter.
—*Shaft* vibration. This is generally not sinusoidal. It normally uses displacement (s) as the measured parameter.

Bearing Housing Vibration

Relevant points are:

—It measures vibration only at the "surface."
—It excludes torsional vibration.
—V_{rms} is normally measured across the frequency range and then distilled down to a single value, i.e.,

$$V_{rms} = \sqrt{\tfrac{1}{2}(\Sigma \text{ amplitudes} \times \text{angular frequences})}$$

Acceptance Levels

Technical standards, and manufacturers' practices, differ in their acceptance levels. General "rule of thumb" acceptance levels are shown in Figs. 5.5. and 5.6.

5.8 Machinery Noise

5.8.1 Principles

Noise is most easily thought of as airborne pressure pulses set up by a vibrating surface source. It is measured by an instrument that detects these pressure changes in the air and then relates this measured sound pressure to an accepted *zero* level.

Fig. 5–5

Machine	V_{rms} in 'mils'/s[in × 10^{-3}/s]	(mm/s)
Precision components and machines – gas turbines, etc.	44	1.12
Helical and epicyclic gearboxes	70	1.8
Spur-gearboxes, turbines	110	2.8
General service pumps	177	4.5
Long-shaft pumps	177–280	4.5–7.1
Diesel engines	280	7.1
Reciprocating large machines	280–440	7.1–11.2

Because a machine produces a mixture of frequencies (termed *broad-band* noise), there is no single noise measurement that will fully describe a noise emission. In practice, two methods used are:

—The *overall noise* level. This is often used as a colloquial term for what is properly described as the *A-weighted sound pressure level.* It incorporates multiple frequencies and weights them according to a formula, which results in the best approximation of the loudness of the noise. This is displayed as a single instrument reading expressed as decibels dB(A).

—*Frequency band* sound pressure level. This involves measuring the sound pressure level in a number of frequency bands. These are arranged in either octave or one-third octave bands in terms of their mid-band frequency. The range of frequencies of interest in measuring machinery noise is from about 30 Hz to 10000 Hz. Note that frequency band sound pressure levels are also expressed in decibels (dB).

The decibel scale itself is a logarithmic scale—a sound pressure level in dB being defined as:

$$\text{dB} = 10 \log_{10} (p_1/p_0)^2$$

where

p_1 = measured sound pressure
p_0 = a reference *zero* pressure level

Noise tests on rotating machines are carried out by defining a "reference surface" and then positioning microphones at locations 3 ft (0.91m) from it (see Fig. 5.7).

Fig. 5–6

Typical balance grades: from International Standard ISO 1940–1

Balance grade	Types of rotor (general examples)
G 1	Grinding machines, tape-recording equipment
G2.5	Turbines, compressors, electric armatures
G6.3	Pump impellers, fans, gears, machine tools
G 16	Cardan shafts, agricultural machinery
G 40	Car wheels, engine crankshafts
G 100	Complete engines for cars and trucks

"Acceptance criteria": from International Standard ISO 10816–1

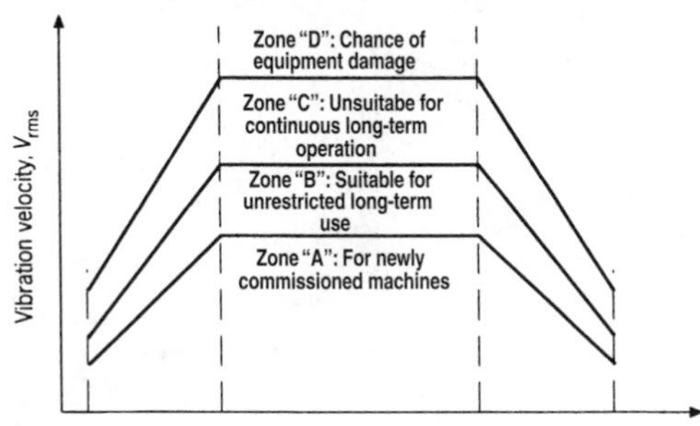

Typical "boundary limits": from International Standard ISO 10816–1

<table>
<tr><th>V_{rms} in 'mils'/s</th><th>(mm/s)</th><th>Class I</th><th>Class II</th><th>Class III</th><th>Class IV</th></tr>
<tr><td>28</td><td>0.71</td><td>A</td><td rowspan="2">A</td><td rowspan="3">A</td><td rowspan="4">A</td></tr>
<tr><td>44</td><td>1.12</td><td rowspan="2">B</td></tr>
<tr><td>70</td><td>1.8</td><td rowspan="2">B</td></tr>
<tr><td>110</td><td>2.8</td><td rowspan="2">C</td><td rowspan="2">B</td></tr>
<tr><td>177</td><td>4.5</td><td rowspan="2">C</td><td rowspan="2">B</td></tr>
<tr><td>280</td><td>7.1</td><td rowspan="3">D</td><td rowspan="2">C</td></tr>
<tr><td>440</td><td>11.2</td><td rowspan="2">D</td><td rowspan="2">C</td></tr>
<tr><td>704</td><td>18</td><td>D</td></tr>
</table>

Class suitability

Class I	Machines < 20.1 hp(≈15kW)
Class II	Machines < 402.1 hp(≈300kW)
Class III	Large machines with rigid foundations
Class IV	Large machines with "soft" foundations

(Note how wide these classes are)

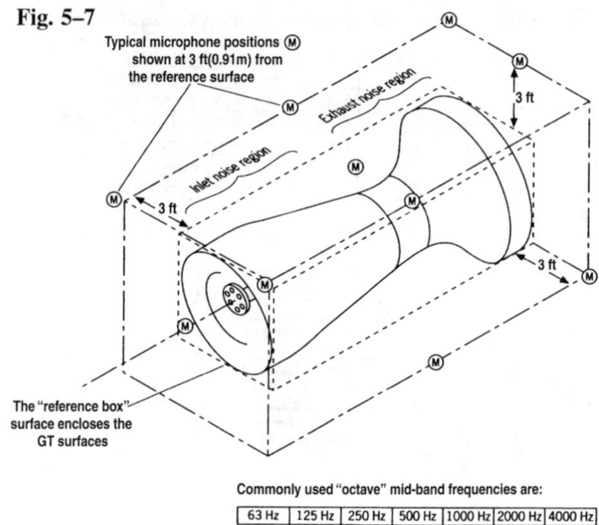

Fig. 5–7

Commonly used "octave" mid-band frequencies are:

63 Hz	125 Hz	250 Hz	500 Hz	1000 Hz	2000 Hz	4000 Hz

5.8.2 Typical Levels

Approximate "rule of thumb" noise levels are given in Table 5.2.

Table 5.2 Typical Noise Levels

Machine/environment	*dB(A)*
A whisper	20
Office noise	50
Noisy factory	90
Large diesel engine	97
Turbocompressor/gas turbine	98

A normal "specification" level is 90 95 dB(A) at 3 ft from the operating equipment. Noisier equipment need an acoustic enclosure. Humans can continue to hear increasing sound levels up to about 120 dB; above this, sound levels cause serious discomfort and long-term damage.

5.9 Useful References

Standards: Balancing

1. API publication 684 Ed 1: *A Tutorial on the API Approach to Rotor Dynamics and Balancing.*
2. SAE ARP 5323: 1988. *Balancing Machines for Gas Turbine Rotors.*

Standards: Vibration

1. ANSI/ASA S2.41:1997. *Mechanical Vibration of Large Rotating Machines With Speed Range from 10 to 200 rev/s.*
2. ANSI/ASA S2.40: 1997. *Mechanical Vibration of Rotating and Reciprocating Machinery—Requirements for Instruments for Measuring Vibration Severity.*

Standards: Noise

1. ANSI/ASA S12.16: 1997. *American National Standard Guidelines for the Specification of Noise from New Machinery.*
2. ANSI/ASA S12.3: 1996. *American National Standard Statistical Methods for Determining and Verifying Stated Noise Emission Values of Machinery and Equipment.*
3. ISO 10494: 1993. *Gas Turbine and Gas Turbine Sets—Measurement of Emitted Airborne Noise—Engineering (Survey Method).*

Section 6

Mechanics of Materials

Table 6.1 Notation

ϵ	Deformation (linear)
$\propto$	Angle of deformation
μ	Poisson's ratio
E	Modulus of elasticity
G	Shear modulus
I	Rectangular moment of area
I_p	Polar moment of area
M	Bending moment
M_t	Twisting (torsion) moment
P	Imposed force
r_c	Radius of curvature
S	Apparent stress
S_c	Ultimate stress (in compression)
S_m	Ultimate stress (in tension)
S_p	Proportional (elastic) limit
S_s or S_v	Shear stress
S_t	Temperature stress
S_y	Yield point
U	Strain energy
V	Shear load
W	Weight of load
w	Distributed load
y	Deflection
Z	Section modulus

6.1 Engineering Structures: Where Are All the Pin Joints?

Much of engineering mechanics is based on the assumption that parts of structures are connected by pin joints. Similarly, members are continually assumed to be "simply-supported" and structural members pretend to be infinitely long, compared with their section thickness. The question is: Do such members really exist?

They are certainly not immediately apparent—look at a bridge or steel tower and you will struggle to find a single joint containing a pin. The structural members will be channels, I-beams, or box sections surrounded by a clutter of plates, gussets, and flanges, not simple beams of nice prismatic section. So where is the relevance of all those clean theories of statics and vector mechanics?

Fortunately, the answer exists already, hidden in two hundred years of engineering experience. Calculations based on simple bending theory, for example, have been validated against actual maximum stresses and deformations experienced in real structures and proved sufficiently accurate (say $\pm$ 10%) to represent reality. Once a factor of safety is introduced, then the simplified calculations are as accurate as they need to be. They are, to all intents and purpose, *correct.*

Simply-supported assumptions work the same way. The complicated-looking supports of a bridge deck do act like simple supports when you consider the length of the beam-like members they are supporting. Equally, the members themselves dissipate stresses induced by constraint from the "real" supports within a short distance from the support, so they *act like* long thin members, even though they may not be.

The design of engineering structures is built around findings like this. They have been proven quantitatively, by using strain-gauges and measuring deflections, and by advanced techniques such as FE analysis and photo-elastic models. Complete structures, airplanes, ships, and buildings have been investigated to demonstrate the validity of taught theories of statics and mechanics. The result is that all these types of structures in the world are designed using equations that are unerringly similar—proof enough of the validity of the theories behind them. Try to improve theoretical techniques, by all mean, but don't ignore what has been found already, including those assumptions about pin joints and simply supported beams.

6.2 Simple Stress and Strain (Deformation)

$$\text{unit stress S} = \frac{\text{load}}{\text{area}} = \frac{\text{P}}{\text{A}}\ \text{lb/in}^2$$

$$\text{Deformation } \epsilon = \frac{\text{change in length}}{\text{original length}} = \text{dl}$$

$$= \text{a ratio therefore no units}$$

Hooke's law:

$$\frac{\text{stress}}{\text{deformation}} = \text{constant} = \text{Young's modulus E lb/in}^2$$

Figure 6.1 shows the general relationship between stress and strain for metallic materials.

Fig. 6–1

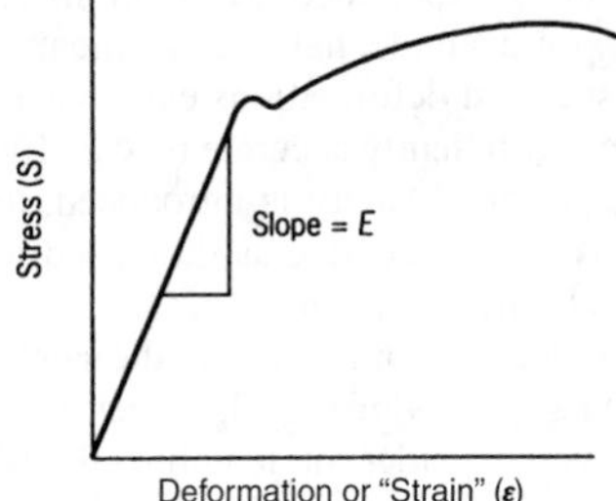

$$\text{Poisson's ratio } (\mu) = \frac{\text{unit lateral deformation (strain)}}{\text{unit longidutinal deformation (strain)}}$$

$$= \frac{\delta d/d}{\delta l/l}$$

ratio, therefore no units (see Fig. 6.2).

Fig. 6–2

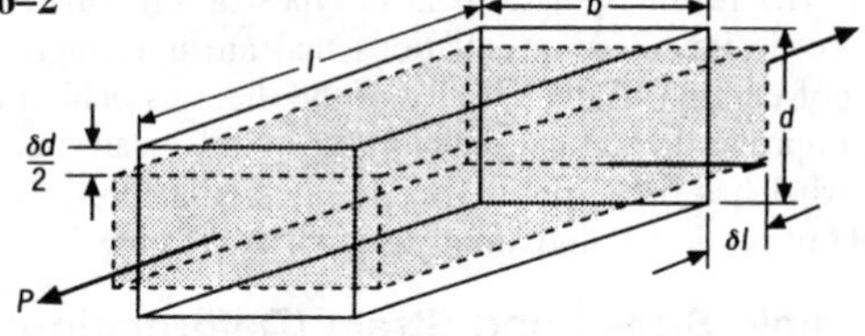

$$\text{Shear stress } (S_v) = \frac{\text{shear load}}{\text{area}} = \frac{V}{A}\text{: units: lb/in}^2$$

Shear deformation ($\propto$) = angle of deformation under shear stress

$$\text{Shear modulus} = \frac{\text{Shear stress}}{\text{Shear strain}} = \frac{S_V}{\propto}$$
$$= \text{constant (G) lb/in}^2$$

(see Fig. 6.3).

Fig. 6–3

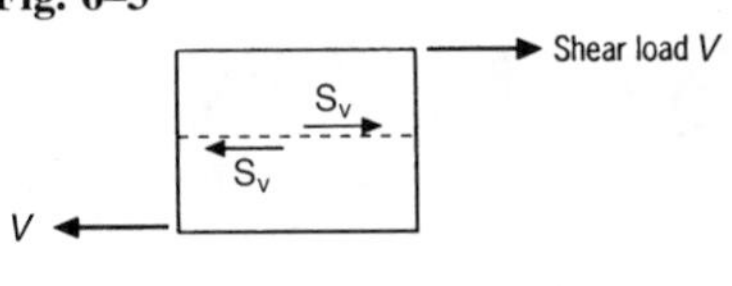

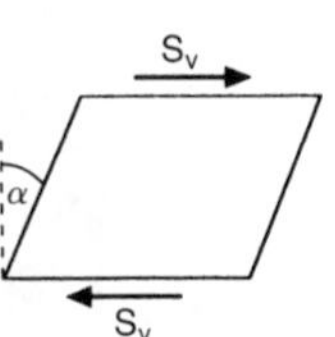

Temperature stress $S_t \cong E\epsilon = Eat$

where

a = linear coefficient of expansion per degree of temperature rise

t = temperature change (Fig. 6.4)

Fig. 6–4

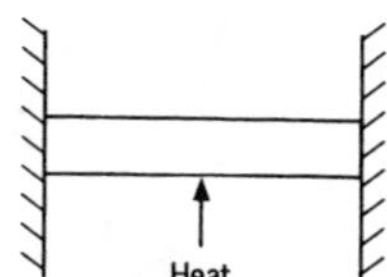

6.3 Simple Elastic Bending (Flexure)

Simple theory of elastic bending is:

$$\frac{M}{I} = \frac{S}{y} = \frac{E}{r_c}$$

M = applied bending moment
I = Rectangular moment of area
r_c = Radius of curvature of neutral axis
E = Elastic modulus
S = Fiber stress due to bending at distance y from neutral axis.

The rectangular moment of area is defined, for any section, as

$I = y^2 dA$

I for common sections is calculated as follows in Fig. 6.5.

Fig. 6–5

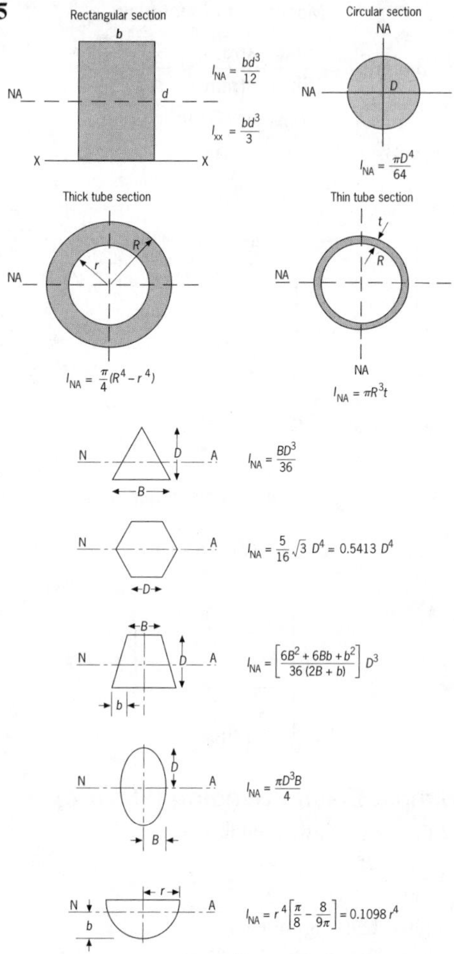

Rectangular section
b
NA
d
X
X
$I_{NA} = \frac{bd^3}{12}$
$I_{xx} = \frac{bd^3}{3}$
Circular section
NA
NA
D
$I_{NA} = \frac{\pi D^4}{64}$
Thick tube section
NA
R
r
$I_{NA} = \frac{\pi}{4}(R^4 - r^4)$
Thin tube section
t
R
NA
NA
$I_{NA} = \pi R^3 t$
N
A
D
B
$I_{NA} = \frac{BD^3}{36}$
N
A
D
$I_{NA} = \frac{5}{16}\sqrt{3}\ D^4 = 0.5413\ D^4$
B
N
A
D
b
$I_{NA} = \left[\frac{6B^2 + 6Bb + b^2}{36\ (2B + b)}\right] D^3$
N
A
D
B
$I_{NA} = \frac{\pi D^3 B}{4}$
r
N
A
b
$I_{NA} = r^4\left[\frac{\pi}{8} - \frac{8}{9\pi}\right] = 0.1098\ r^4$

I about another axix (XX) can be found using the parallel axis theorem:

Area A
$I_{XX} = I_{NA} + Ad^2$
d
X
X

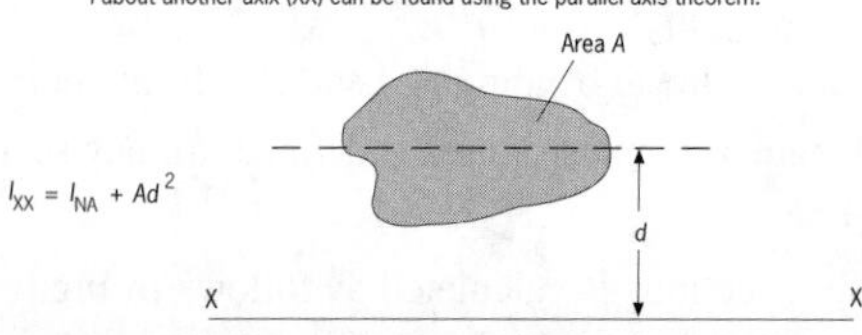

Fig. 6–5 (cont.)

Steelwork sections

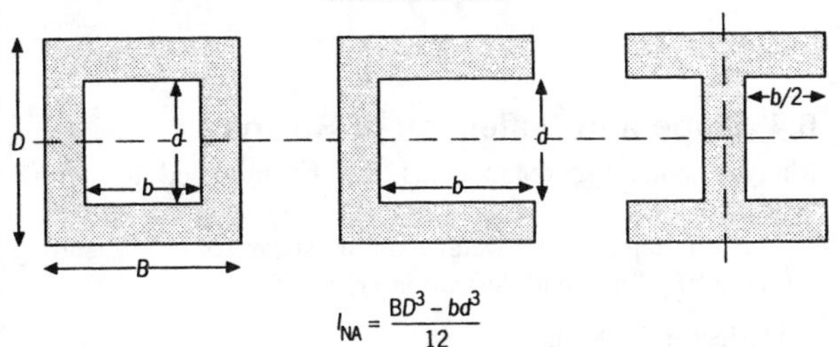

$$I_{NA} = \frac{BD^3 - bd^3}{12}$$

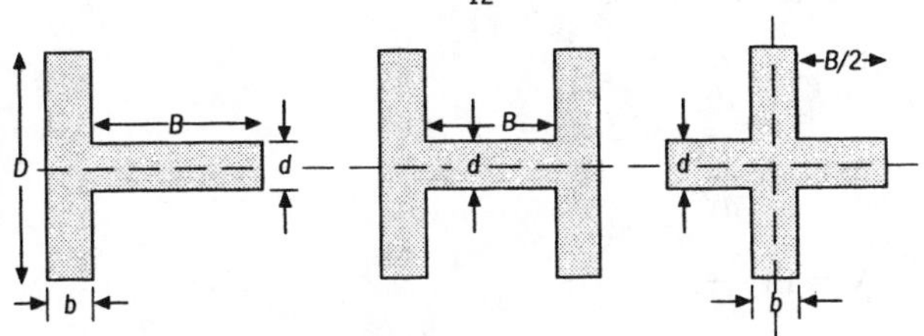

$$I_{NA} = \frac{bD^3 + Bd^3}{12}$$

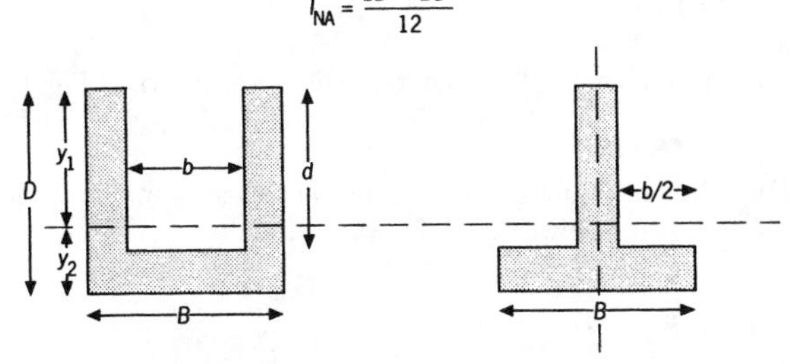

$$y_1 = \frac{BD^2 - bd^2}{2(BD - bd)}$$

$$I = \frac{(BD^2 - bd^2)^2 - 4BDbd(D - d)^2}{12(BD - bd)}$$

$$y_2 = \frac{BD^2 - 2bdD + bd^2}{2(BD - bd)}$$

Section modulus Z is defined as

$$Z = \frac{I}{y}$$

6.4 Slope and Deflection of Beams

Many engineering components can be modeled as simple beams.

The relationships between load W, shear force V, bending moment M, slope, and deflection (y) are

Deflection = δ (or y)

$$\text{Slope} = \frac{dy}{dx}$$

$$M = EI\,\frac{d^2y}{dx^2}$$

$$V = EI\,\frac{d^3y}{dx^2}$$

$$W = EI\,\frac{d^4y}{dx^4}$$

Values for common beam configurations are shown in Fig. 6.6.

6.5 Torsion

For solid or hollow shafts of uniform cross-section, the torsion formula is (see Figs. 6.7 and 6.8):

$$\frac{T}{I_p} = \frac{S_v}{R} = \frac{G\theta}{l}$$

T = torque applied (ft.lb)
I_p = polar second moment of area (in^4)
S_v = shear stress (lb/in^2)
R = radius (in)
G = Shear modulus (lb/in^2)
θ = angle of twist (rad)
l = length

For solid shafts: $I_p = \dfrac{\pi D^4}{32}$

For hollow shafts: $I_p = \dfrac{\pi(D^4 - d^4)}{32}$

Fig. 6–6

Conditions of support and loading	Bending moment (maximum)	Shearing force (maximum)	Safe load W	Deflection (maximum)
 W; L	WL	W	$\frac{M}{L}$	$\frac{WL^3}{3EI}$
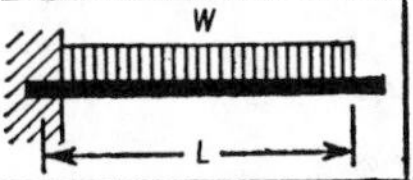 W; L	$\frac{WL}{2}$	W	$\frac{2M}{L}$	$\frac{WL^3}{8EI}$
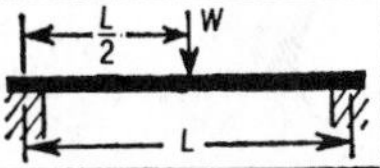 $\frac{L}{2}$; W; L	$\frac{WL}{4}$	$\frac{W}{2}$	$\frac{4M}{L}$	$\frac{WL^3}{48EI}$
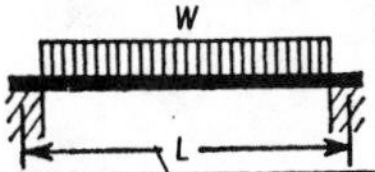 W; L	$\frac{WL}{8}$	$\frac{W}{2}$	$\frac{8M}{L}$	$\frac{5WL^3}{384EI}$
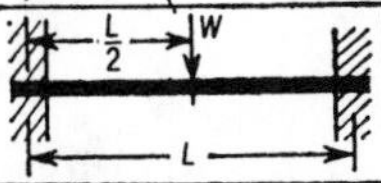 $\frac{L}{2}$; W; L	$\frac{WL}{8}$	$\frac{W}{2}$	$\frac{8M}{L}$	$\frac{WL^3}{192EI}$
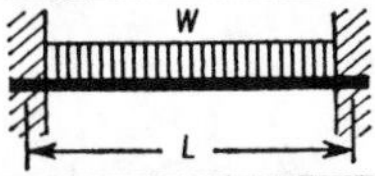 W; L	$\frac{WL}{12}$	$\frac{W}{2}$	$\frac{12M}{L}$	$\frac{WL^3}{384EI}$
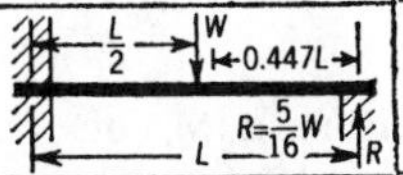 $\frac{L}{2}$; W; 0.447L; $R=\frac{5}{16}W$; L; R	$\frac{3WL}{16}$	$\frac{11W}{16}$	$\frac{16M}{3L}$	$\frac{WL^3}{107EI}$
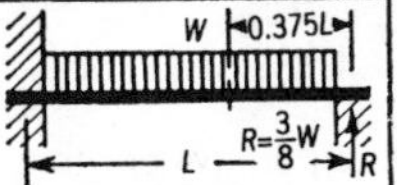 W; 0.375L; $R=\frac{3}{8}W$; L; R	$\frac{WL}{8}$	$\frac{5W}{8}$	$\frac{8M}{L}$	$\frac{WL^3}{187EI}$

For thin-walled hollow shafts: $I_p \cong 2\pi r^3 t$

Where

r = mean radius of shaft wall
t = wall thickness

Shaft under combined bending moment, *M*, and torque, *T*

Fig. 6–7

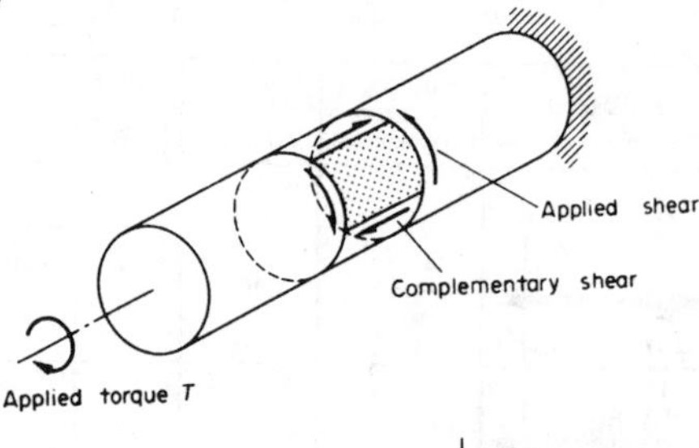

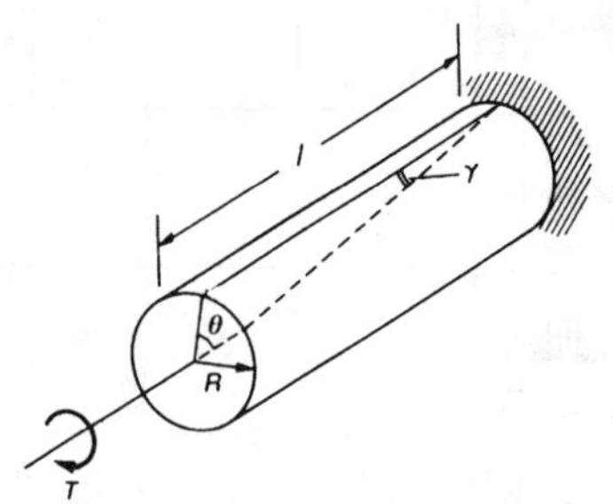

From bending: $S = \dfrac{MD}{2I}$

From torsion: $S_v = \dfrac{TD}{2I_p}$

This results in an "equivalent" bending moment (M_e) of:

$$M_e = \frac{1}{2}\left[\sqrt{(M^2 + T^2)}\right]$$

A similar approach can be used to give an equivalent torque (T_e)

$$T_e = \sqrt{(M^2 + T^2)}$$

6.6 Thin Cylinders

Most pressure vessels have a diameter:wall thickness ratio of > 20 and can be modeled using thin cylinder assumptions. The basic equations form the basis of all pressure vessel codes and standards.

Fig. 6–8

TORSION FORMULAE

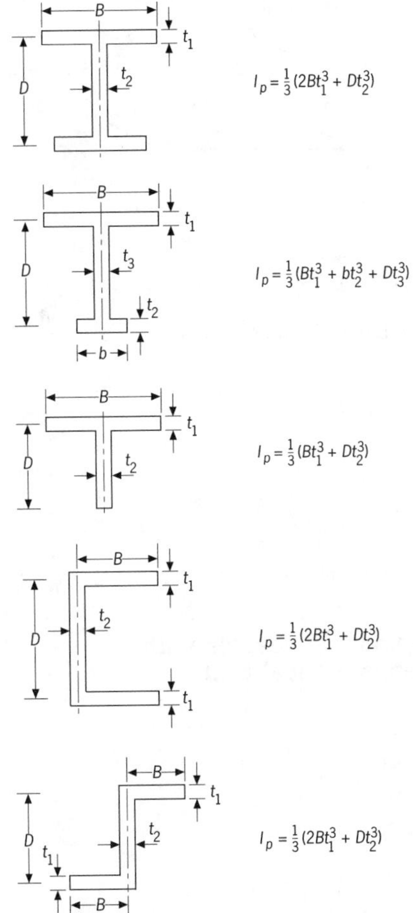

The polar second motion of area (I_p) in^4 is a measure of the stiffness of a member in pure twisting

Fig. 6–9

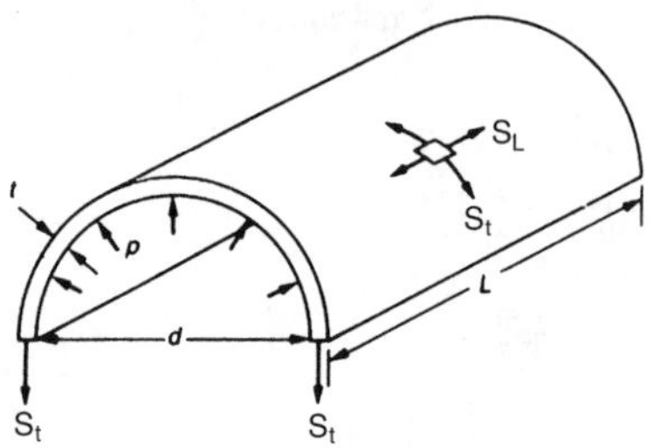

Basic equations are (see Fig. 6.9)

Circumferential (Hoop) stress $S_t = \frac{pd}{2t}$

Hoop strain $\epsilon_t = \frac{I}{E}(S_t - \mu S_L)$

Longitudinal (axial) stress $S_L = \frac{pd}{4t}$

Longitudinal strain $\varepsilon_L = \frac{1}{E}(S_L - \mu S_t)$

6.7 Cylindrical Vessels With Hemispherical Ends

(see Fig. 6.10)

For the cylinder:

$$S_C = \frac{pd}{2t_c} \text{ and } S_{LC} = \frac{pd}{2t_c}$$

Hoop strain:

$$\epsilon_{HC} = \frac{1}{E}(S_{tC} - \mu S_{LC})$$

For the hemispherical ends:

$$S_{ts} = \frac{pd}{4t_s} \text{ and } \epsilon_{ts} = \frac{pd}{4t_sE}(1 - \mu)$$

Fig. 6–10

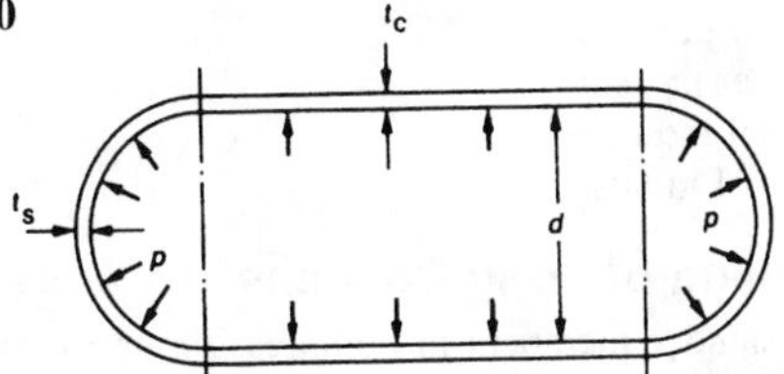

The differences in deformation strain produce *discontinuity stress* at a vessel head/shell joint.

6.8 Thick Cylinders

Components such as hydraulic rams and boiler headers are designed using thick cylinder assumptions. Hoop and radial stresses vary through the walls, giving rise to the Lamé equations (see Fig. 6.11)

Fig. 6–11

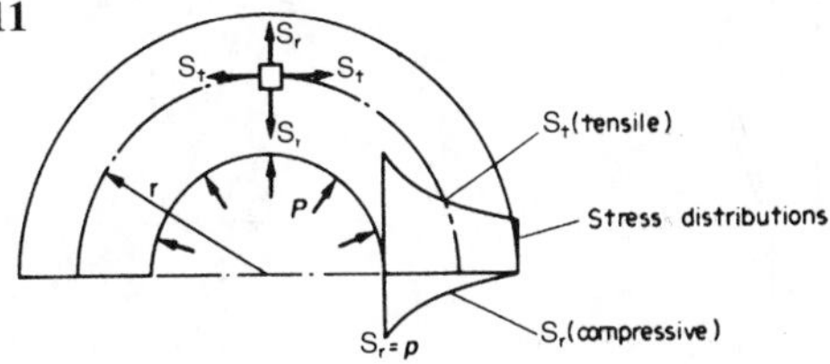

$$S = A + \frac{B}{r^2} \text{ and } S_r = A - \frac{B}{r^2}$$

Where A and B are "Lamé" constants

$$\varepsilon_t = \frac{S_t}{E} - \frac{\mu S_r}{E} - \frac{\mu S_L}{E}$$

$$\varepsilon_L = \frac{S_L}{E} - \frac{\mu S_r}{E} - \frac{\mu S_t}{E}$$

Lamé constant (A) is given by:

$$A = \frac{P_1 R_1{}^2 - P_2 R_2{}^2}{R_2{}^2 - R_1{}^2}$$

P_1 = Internal pressure
P_2 = External pressure
R_1 = Internal radius
R_2 = External radius

6.9 Buckling of Long Columns

Long and slender members in compression are termed "long columns." They fail by buckling before reaching their true compressive yield strength. Buckling load W_b depends on the loading case (see Fig. 6.12).

Fig. 6–12

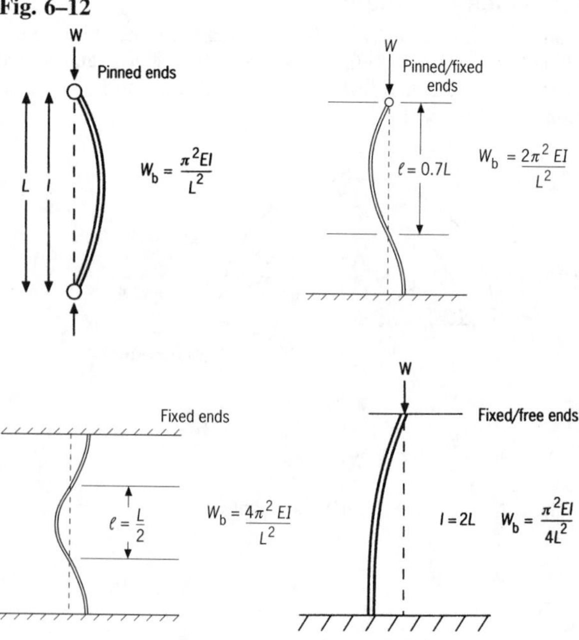

The *equivalent length, l,* of the strut is the length of a single "bow" in the deflected condition.

6.10 Flat Circular Plates

Many parts of engineering assemblies can be analyzed by approximating them to flat circular plates or annular rings. The general equation governing slopes and deflections is (see Fig. 6.13)

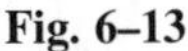

Fig. 6–13

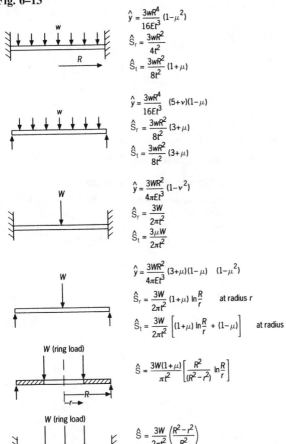

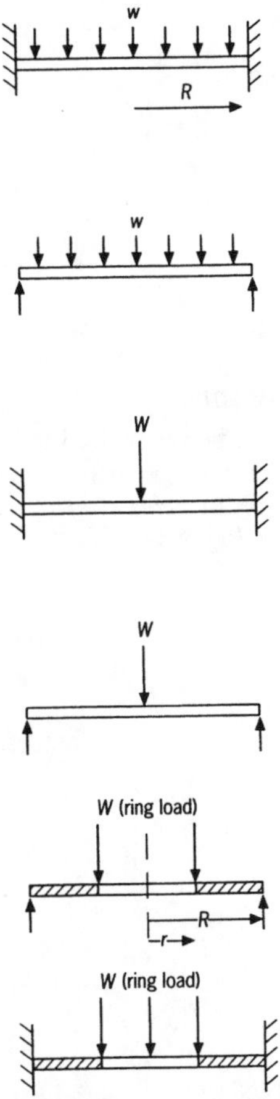

$$\hat{y} = \frac{3wR^4}{16Et^3}(1-\mu^2)$$
$$\hat{S}_r = \frac{3wR^2}{4t^2}$$
$$\hat{S}_t = \frac{3wR^2}{8t^2}(1+\mu)$$

$$\hat{y} = \frac{3wR^4}{16Et^3}\quad(5+\nu)(1-\mu)$$
$$\hat{S}_r = \frac{3wR^2}{8t^2}(3+\mu)$$
$$\hat{S}_t = \frac{3wR^2}{8t^2}(3+\mu)$$

$$\hat{y} = \frac{3WR^2}{4\pi Et^3}(1-\nu^2)$$
$$\hat{S}_r = \frac{3W}{2\pi t^2}$$
$$\hat{S}_t = \frac{3\mu W}{2\pi t^2}$$

$$\hat{y} = \frac{3WR^2}{4\pi Et^3}(3+\mu)(1-\mu)\quad(1-\mu^2)$$
$$\hat{S}_r = \frac{3W}{2\pi t^2}(1+\mu)\ln\frac{R}{r}\qquad \text{at radius } r$$
$$\hat{S}_t = \frac{3W}{2\pi t^2}\left[(1+\mu)\ln\frac{R}{r} + (1-\mu)\right]\qquad \text{at radius } r$$

$$\hat{S} = \frac{3W(1+\mu)}{\pi t^2}\left[\frac{R^2}{(R^2-r^2)}\ln\frac{R}{r}\right]$$

$$\hat{S} = \frac{3W}{2\pi t^2}\left(\frac{R^2-r^2}{R^2}\right)$$

$$\frac{d}{dr}\left[\frac{1}{r}\frac{d}{dr}\left(r\frac{dy}{dr}\right)\right] = \frac{W}{D}$$

Where $D = \dfrac{Et^3}{12(1 - \mu^2)}$

$\hat{y}$ = maximum deflection

$\dfrac{dy}{dr}$ = slope

W = applied load

t = thickness

D = flexural stiffness

E = Young's modulus

$\hat{S}_r$ = maximum radial stress

$\hat{S}_t$ = maximum tangential stress

6.11 Stress Concentration Factors

The effective stress in a component can be raised well above its expected levels owing to the existence of geometrical features causing stress concentrations under dynamic elastic conditions. Typical factors are as shown in Fig. 6.14.

Fig. 6–14a

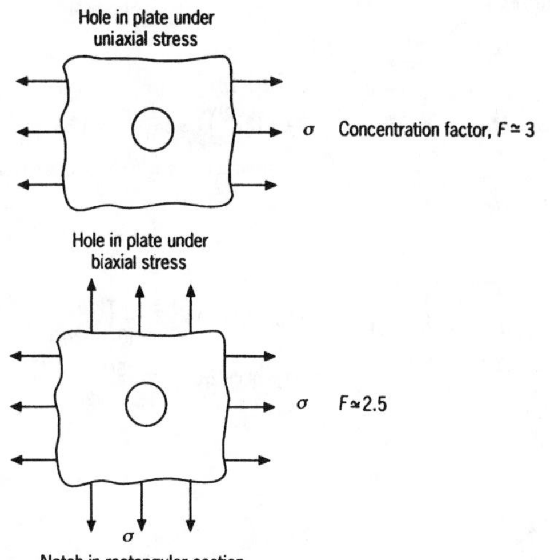

Notch in rectangular section under bending

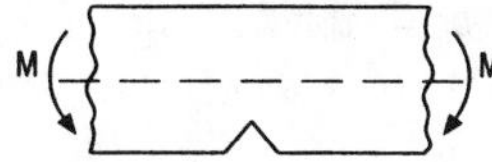

F≃4–7, depending on sharpness of the notch

Fillet radius in shaft under bending

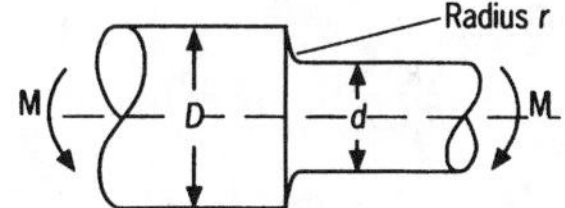

D/d \ r/d	0.05	0.15	0.2	0.3
1.0	1.8	1.5	1.4	1.3
1.5	2.2	1.6	1.4	1.3
3.0	2.6	1.7	1.5	1.4

Approximate value of stress concentration factor, *F*

Fig. 6–14b

APPROXIMATE STRESS CONCENTRATION FACTORS (Elastic Stresses)

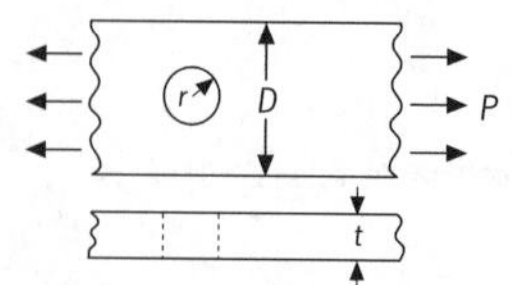

Hole in plate under uni-axial stress

$$\hat{S} \simeq F\left[\frac{P}{t(D-2r)}\right]$$

$$F \simeq 3 - 3.13\left(\frac{2r}{D}\right) + 3.66\left(\frac{2r}{D}\right)^2 - 1.53\left(\frac{2r}{D}\right)^3$$

The maximum stress is at the edge of the hole

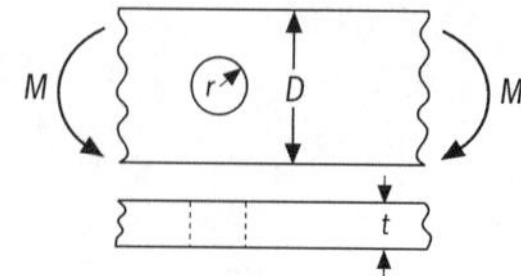

Hole in plate under in-plane bending

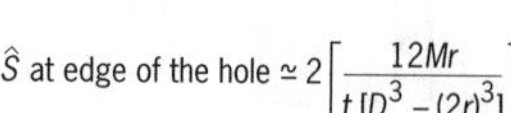

$$\hat{S} \text{ at edge of the hole} \simeq 2\left[\frac{12Mr}{t[D^3 - (2r)^3]}\right]$$

$$\hat{S} \text{ at edge of the plate} \simeq \frac{6MD}{t[D^3 - (2r)^3]}$$

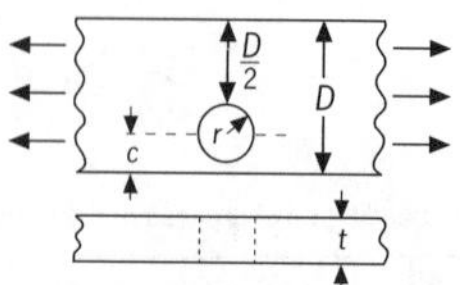

Off-center hole in plate under uni-axial stress

$$\hat{S} = FS_{nominal}$$

$$S_{nominal} \simeq \frac{P\sqrt{1-(r/c)^2}\,[1-(c/D)]}{Dt[1-(r/c)]\,[1-(c/D)]\,\left[2-\sqrt{1-(r/\right.}}$$

$$F \simeq 3 - 3.13\,(r/c) + 3.66\,(r/c)^2 - 1.53\,(r/c)^3$$

Elliptical hole in plate under uni-axial stress

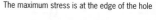

$$\hat{S} = FS_{nominal}$$

$$S_{nominal} = \frac{P}{t(D-2a)}$$

$$F \simeq F_1 + F_2\left(\frac{2a}{D}\right) + F_3\left(\frac{2a}{D}\right)^2 + F_4\left(\frac{2a}{D}\right)^3$$

for $0.5 \le \frac{a}{b} \le 10$

$$F_1 \simeq 1 + \sqrt{\frac{a}{b}} + \frac{2a}{b}$$

$$F_2 \simeq -0.351 - 0.021\sqrt{\frac{a}{b}} - \frac{2.483a}{b}$$

$$F_3 \simeq -3.621 - 5.183\sqrt{\frac{a}{b}} + \frac{4.494a}{b}$$

$$F_4 \simeq -2.27 + 5.2\sqrt{\frac{a}{b}} - \frac{4a}{b}$$

Fig. 6–14c

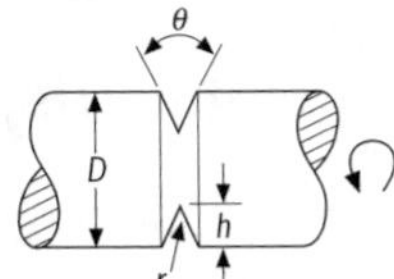

V-notch in circular shaft under torsion

F_v = stress concentration factor for V-notch

$$F_v \simeq F_u - \left[0.02 + 0.14\left(\frac{\theta}{135}\right)^2\right](F_u - 1)\,F_u$$

where F_u = stress concentration factor for U-notch in torsion

for $\frac{r}{D-2h} \leq 0.01$ and $\theta \leq 135^\circ$

U-notch in circular shaft under axial tension

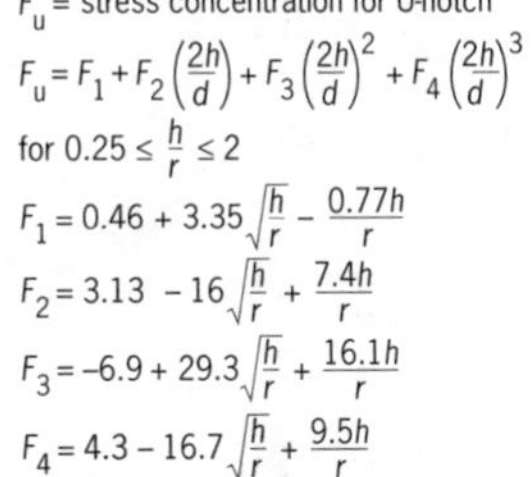

F_u = stress concentration for U-notch

$$F_u = F_1 + F_2\left(\frac{2h}{d}\right) + F_3\left(\frac{2h}{d}\right)^2 + F_4\left(\frac{2h}{d}\right)^3$$

for $0.25 \leq \frac{h}{r} \leq 2$

$$F_1 = 0.46 + 3.35\sqrt{\frac{h}{r}} - \frac{0.77h}{r}$$

$$F_2 = 3.13 - 16\sqrt{\frac{h}{r}} + \frac{7.4h}{r}$$

$$F_3 = -6.9 + 29.3\sqrt{\frac{h}{r}} + \frac{16.1h}{r}$$

$$F_4 = 4.3 - 16.7\sqrt{\frac{h}{r}} + \frac{9.5h}{r}$$

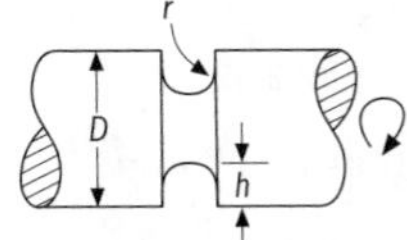

U-notch in circular shaft under torsion

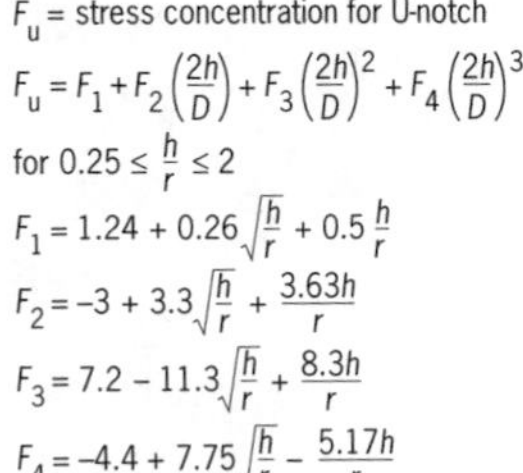

F_u = stress concentration for U-notch

$$F_u = F_1 + F_2\left(\frac{2h}{D}\right) + F_3\left(\frac{2h}{D}\right)^2 + F_4\left(\frac{2h}{D}\right)^3$$

for $0.25 \leq \frac{h}{r} \leq 2$

$$F_1 = 1.24 + 0.26\sqrt{\frac{h}{r}} + 0.5\,\frac{h}{r}$$

$$F_2 = -3 + 3.3\sqrt{\frac{h}{r}} + \frac{3.63h}{r}$$

$$F_3 = 7.2 - 11.3\sqrt{\frac{h}{r}} + \frac{8.3h}{r}$$

$$F_4 = -4.4 + 7.75\sqrt{\frac{h}{r}} - \frac{5.17h}{r}$$

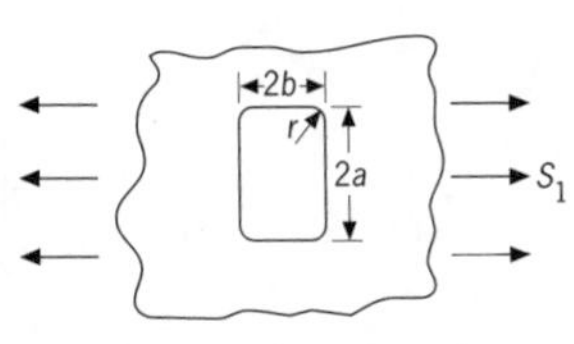

Rectangular hole with round corners in "infinite" plate under uniaxial stress

$$\hat{S} = FS_1$$

$$F \simeq F_1 + F_2\left(\frac{b}{a}\right) + F_3\left(\frac{b}{a}\right)^2 + F_4\left(\frac{b}{a}\right)^3$$

for $0.2 \leq \frac{r}{b} \leq 1$ and $0.3 \leq \frac{b}{a} \leq 1$

$$F_1 \simeq 14.8 - 15.8\sqrt{\frac{r}{b}} + \frac{8.15r}{b}$$

$$F_2 = -11.2 - 9.7\sqrt{\frac{r}{b}} + \frac{9.6r}{b}$$

$$F_3 = 0.2 + 38.6\sqrt{\frac{r}{b}} - \frac{27.4r}{b}$$

$$F_4 = 3.2 - 23\sqrt{\frac{r}{b}} + \frac{15.5r}{b}$$

Fig. 6–14d

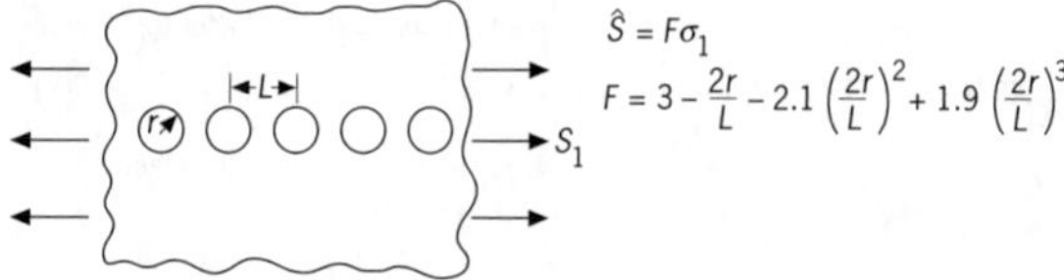

$$\hat{S} = F\sigma_1$$

$$F = 3 - \frac{2r}{L} - 2.1\left(\frac{2r}{L}\right)^2 + 1.9\left(\frac{2r}{L}\right)^3$$

Row of circular holes in "infinite" plate under uniaxial stress

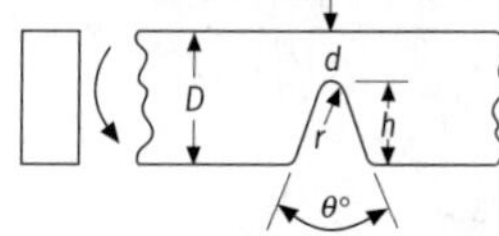

V-notch in rectangular section under bending

F_V = Stress concentration factor for V-notch

$$F_V = 1.11\ F_u - \left[0.03 + 0.11\left(\frac{\theta^\circ}{150}\right)^4\right] F_u^2$$

where F_u = Stress concentration factor for U-notch

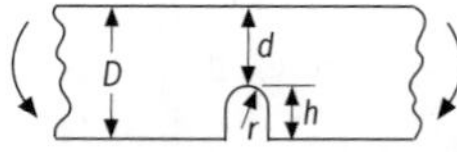

U-notch in rectangular section under bending

F_u = Stress concentration factor for u-notch

$$F_u = F_1 + F_2\left(\frac{h}{d}\right) + F_3\left(\frac{h}{d}\right)^2 + F_4\left(\frac{h}{d}\right)^3$$

for $0.5 \leq \frac{h}{r} \leq 4$

$$F_1 \simeq 0.72 + 2.4\sqrt{\frac{h}{r}} - \frac{0.13h}{r}$$

$$F_2 \simeq 1.98 - 11.5\sqrt{\frac{h}{r}} + \frac{2.2h}{r}$$

$$F_3 \simeq -4.4 + 18.75\sqrt{\frac{h}{r}} + \frac{4.6h}{r}$$

$$F_4 \simeq 2.7 - 9.7\sqrt{\frac{h}{r}} + \frac{2.5h}{r}$$

6.12 Types of Stress Loading

Stress conditions in engineering components are rarely static: They often vary with time. Fig. 6.15 shows the four main classifications of stress loading.

Fig. 6–15

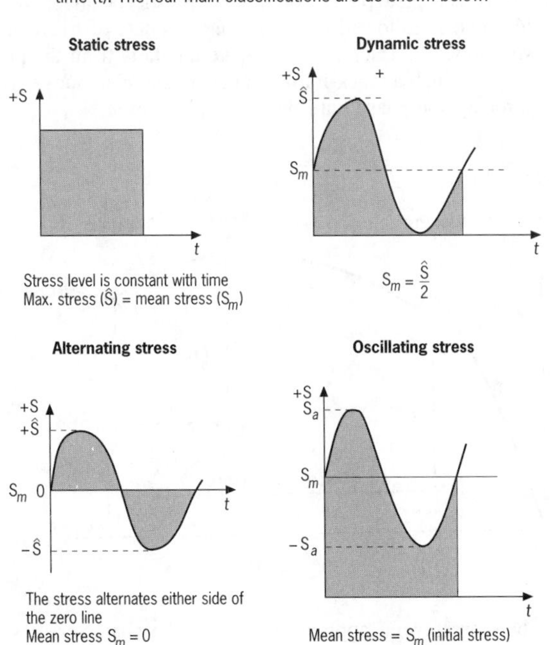
TYPES OF STRESS LOADING
Stresses in engineering components are rarely static – they often vary with time (t). The four main classifications are as shown below:
Static stress
+S
t
Stress level is constant with time
Max. stress (Ŝ) = mean stress (S_m)
Dynamic stress
+S
+
Ŝ
S_m
t
$S_m = \frac{\hat{S}}{2}$
Alternating stress
+S
+Ŝ
S_m 0
t
–Ŝ
The stress alternates either side of the zero line
Mean stress $S_m = 0$
Oscillating stress
+S
S_a
S_m
$-S_a$
t
Mean stress = S_m (initial stress)
Max. stress, $\hat{S} = S_m + S_a$

Section 7

Material Failure

7.1 How Materials Fail

There is no single, universally accepted explanation for the way that materials (particularly metals) fail. Figure 7.1 shows the generally accepted phases of failure. Elastic behavior, up to yield point, is followed by increasing amounts of irreversible plastic flow. The fracture of the material starts from the point in time at which a crack initiation occurs and continues during the propagation phase until the material breaks.

Fig. 7–1

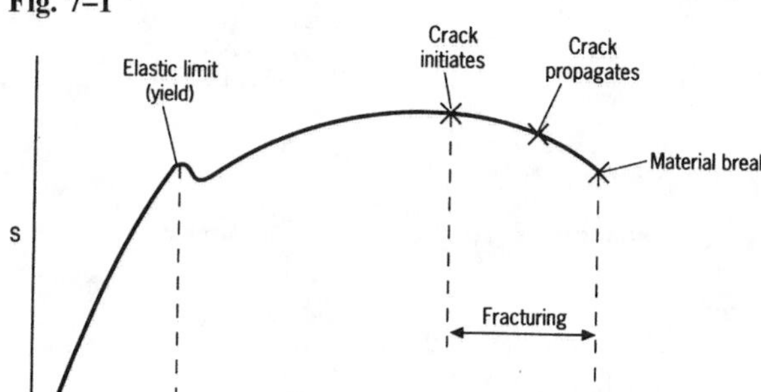

There are several approaches to both the characteristics of the original material and the way the material behaves at a crack tip (see Fig. 7.2). Two of the more common ones are:

—The linear elastic fracture mechanics (LEFM) approach with its related concept of fracture toughness (K_{1c}) parameter (a material property).

—Fully plastic behavior at the crack tip, i.e., "plastic collapse" approach.

A useful standard is ASTM E399.

Fig. 7–2

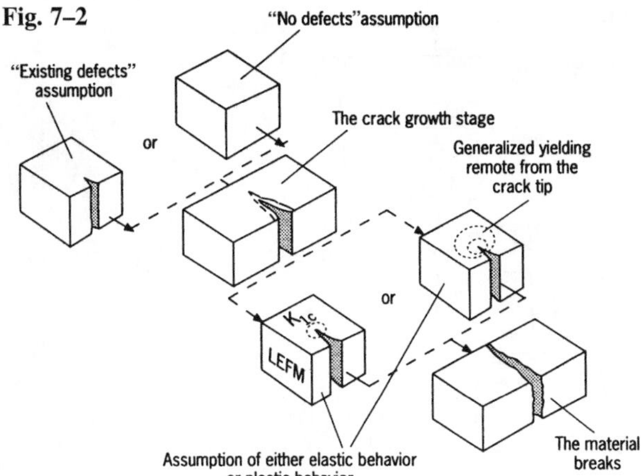

7.2 LEFM Method

This is based on the "fast fracture" equation:

$$K_{1c} = K_1 \equiv YS\sqrt{\pi a}$$

K_{1c} = plane strain fracture toughness
K_1 = stress intensity factor
a = crack length
Y = dimensionless factor based on geometry
S = stress level

Typical Y values used are shown in Fig. 7.3.

7.3 Multi-Axis Stress States

When stress is not uniaxial (as in many real components), yielding is governed by a combination of various stress components acting together. There are several different "approaches" as to how this happens.

Fig. 7–3

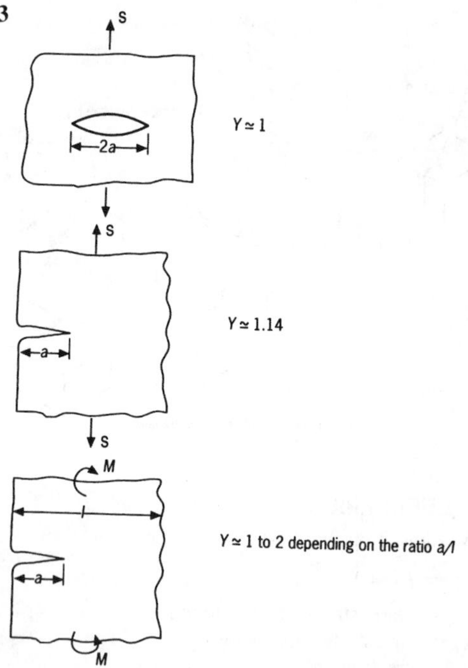

7.3.1 Von Mises Criterion (or "Distortion Energy" Theory)

This states that yielding will take place when

$$\frac{1}{2^{1/2}}\left[(S_1 - S_2)^2 + (S_2 - S_3)^2 + (S_3 - S_1)^2\right]^{1/2} = \pm S_y$$

where S_1, S_2, S_3 are the principal stresses at a point in a component.

It is a useful theory for ductile metals.

7.3.2 Tresca Criterion (or Maximum Shear Stress Theory)

$$\frac{(S_1 - S_2)}{2} \text{ or } \frac{(S_2 - S_3)}{2} \text{ or } \frac{(S_3 - S_1)}{2} = \pm \frac{S_y}{2}$$

This is also a useful theory for ductile materials.

7.3.3 Maximum Principal Stress Theory

This is a simpler theory, which is a useful approximation for brittle metals.

The material fails when

$$S_1 \text{ or } S_2 \text{ or } S_3 = \pm S_y$$

7.4 Fatigue

Ductile materials can fail at stresses significantly less than their rated yield strength if they are subject to fatigue loadings. Fatigue data are displayed graphically on a S–N curve (see Fig. 7.4). Some materials exhibit a "fatigue limit," representing the stress at which the material can be subjected to (in theory) an infinite number of cycles without exhibiting any fatigue effects. This fatigue limit is influenced by the size and surface finish of the specimen, as well as the material's properties.

Fig. 7–4

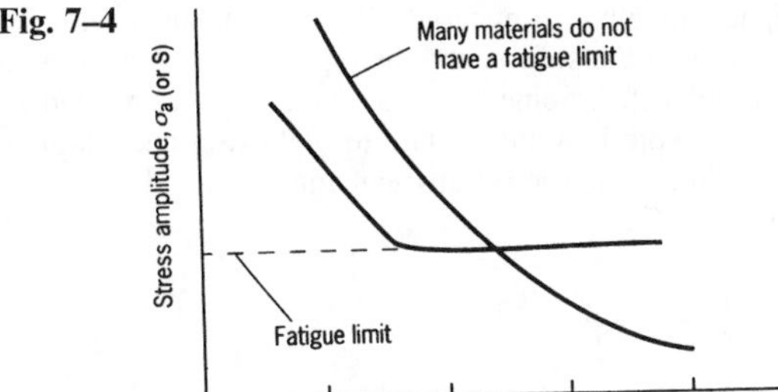

Characteristics of fatigue failures are:

—Visible crack-arrest and "beach mark" lines on the fracture face.

—Striations (visible under magnification)—these are the result of deformation during individual stress cycles.

—An initiation point such as a crack, defect, or inclusion, normally on the surface of the material.

7.4.1 Typical Fatigue Limits

Table 7.1 Typical Fatigue Limits

Material	*Tensile strength* $(S_m) \times$ *1000 lb/in²*	*Approximate fatigue limit* $\times$ *1000 lb/in²*
Low carbon steel	50–140	20–70
Cast iron	20–45	6–16
Cast Steel	60–75	23–30
Titanium	85–90	40–45
Aluminum	18–35	5–10
Brass	25–75	7–19
Copper	30–45	11–15

7.4.2 Fatigue Strength—Rules of Thumb

The fatigue strength of a material varies significantly with the size and shape of the section and the type of fatigue stresses to which it is subjected. Some "rules of thumb" values are shown in Table 7.2. Note how they relate to yield (S_y) and ultimate (S_m) values in pure tension. European equivalent (SI) units are shown in Table 7.3.

Table 7.2 Fatigue Strength: Rules of Thumb

	Bending			*Tension*		*Torsion*		
	$S_{w(b)}$	$S_{a(b)}$	$S_{y(b)}$	S_w	S_a	$S_{sw(t)}$	$S_{sa(t)}$	$S_{sy(t)}$
Steel (structural)	$0.5S_m$	$0.75S_m$	$1.5\ S_y$	$0.45\ S_m$	$0.59\ S_m$	$0.35\ S_m$	$0.38\ S_m$	$0.7\ S_m$
Steel (hardened and tempered)	$0.45S_m$	$0.77S_m$	$1.4\ S_y$	$0.4\ S_m$	$0.69\ S_m$	$0.3\ S_m$	$0.5\ S_m$	$0.7\ S_y$
Cast Iron	$0.38S_m$	$0.68S_m$	—	$0.25\ S_m$	$0.4\ S_m$	$0.35\ S_m$	$0.56\ S_m$	—

$S_{w(b)}$: Fatigue strength under alternating stress (bending).
$S_{a(b)}$: Fatigue strength under fluctuating stress (bending).
$S_{y(b)}$: Yield point (bending).
S_w: Fatigue strength under alternating stress (tension).
S_a: Fatigue strength under fluctuating stress (tension).
S_y: Yield point (tension).
$S_{sw(t)}$: Fatigue strength under alternating stress (torsion).
$S_{sa(t)}$: Fatigue strength under fluctuating stress (torsion).
$S_{sy(t)}$: Yield point (torsion).

Material strength definitions and equivalent units in use in Europe are as shown:

Table 7.3 European Equivalents

	Yield Strength	*Ultimate Tensile Strength*	*Modulus*
USCS	F_{ty}(ksi) or S_y (lb/in^2)	F_{tu}(ksi) or S_m (lb/in^2)	E(lb/in^2 × 10^6)
SI/European	R_e(MN/m^2)	R_m(MN/m^2)	E(N/m^2 × 10^9)
Conversions are:			
1ksi = 1000psi = 6.89MPa = 6.89MN/m^2 = 6.89N/mm^2			

7.5 Margins of Safety

Margins of safety play a part in all aspects of engineering design (see Table 7.4). For safety-critical items such as pressure vessels and cranes, margins are specified in the design codes. In other equipment it is left to established practice and designers' preference. The overall margin of safety in a design can be thought of as being made up of three parts.

- The S_y/S_m ratio;
- The nature of the working load condition, i.e., static, fluctuating, uniform, etc. (see Chapter 6);
- Unpredictable variations, such as accidental overload.

Table 7.4 Typical Margins of Safety

Equipment	*Margins*
Pressure vessels	5–6
Heavy-duty shafting	10–12
Structural steelwork (buildings)	4–6
Structural steelwork (bridges)	5–7
Engine components	6–8
Turbine components (static)	6–8
Turbine components (rotating)	2–3
Airplane components	1.5–2.5
Wire ropes	8–9
Lifting equipment (hooks, etc.)	8–9

Design margins of safety are mentioned in many published technical standards.

Section 8

Thermodynamics and Cycles

Table 8.1 Notation—Thermodynamics

A	Area
As	Surface area
c	Specific heat
c_p	Specific heat (constant pressure)
c_v	Specific heat (constant volume)
D	Diameter
E	Total "system" energy
e	"System" energy per unit mass
Fo	Fourier number
G	Irradiation
Gr	Grashot number
Gz	Graetz number
h	Convection heat transfer coefficient
	Planck's constant
h_{fg}	Enthalpy of vaporization
h_m	Convection mass transfer coefficient
h_{rad}	Radiation heat transfer coefficient
J	Radiosity
k	Thermal conductivity
	Boltzman's constant
M(m)	Mass
Nu	Nusselt number
P	Pitch of a tube-bank
Pe	Peclet number (Re, Pm)
Pr	Prandtl number
p	Pressure
Q	Heat transfer
q	Rate of heat transfer
q	Rate of energy generation per unit volume
R	Ideal gas constant
R_u	Universal gas constant
Re	Reynolds number
R_f	Fouling factor

Table 8.1 Notation—Thermodynamics (cont.)

r	Radius of cylinder or sphere
r, ϕ, z	Cylindrical co-ordinates
r, θ, ϕ	Spherical co-ordinates
St	Stanton number
T	Temperature
t	Time
u	Internal energy
U	Overall heat transfer coefficient
V	Volume
v	Specific volume
x, y, z	Rectangular co-ordinates
α	Thermal diffusivity
β	Volumetric thermal expansion coefficient
δ	Hydrodynamic boundary layer thickness
δ_t	Thermal boundary layer thickness
ϵ	Emissivity
ϵ_f	Fin effectiveness
η_f	Fin efficiency
(Δt)	Temperature difference
K	Absorption coefficient
λ	Wavelength
μ	Dynamic viscosity
ν	Kinematic viscosity
ρ	Density
ρ	Reflectivity
σ	Stefan-Boltzman constant
φ	Stream function
τ	Shear stress

8.1 Basic Thermodynamic Laws

The basic laws of thermodynamics govern the design and operation of engineering machines. The most important principles are those concerned with the conversion of heat energy from available sources such as fuels into useful work.

8.1.1 The First Law

The first law of thermodynamics is merely a specific way to express the principle of conservation of energy. It says, effectively, that heat and work are two mutually convertible forms of energy. So:

heat in = work out

or, in symbols

$\Sigma dQ = \Sigma dW$ (over a complete cycle)

This leads to the non-flow energy equation

$dQ = du + dW$

where u = internal energy.

8.1.2 The Second Law

This can be expressed in several ways:

—Heat flows from hot to cold, not cold to hot.
—In a thermodynamic cycle, gross heat supplied must exceed the net work done—so some heat has to be *rejected* if the cycle is to work.
—A working cycle must have a heat supply and a heat sink.
—The thermal efficiency of a heat engine must always be less than 100 percent.

The two laws point toward the general representation of a heat engine as shown (see Fig. 8.1).

Fig. 8–1

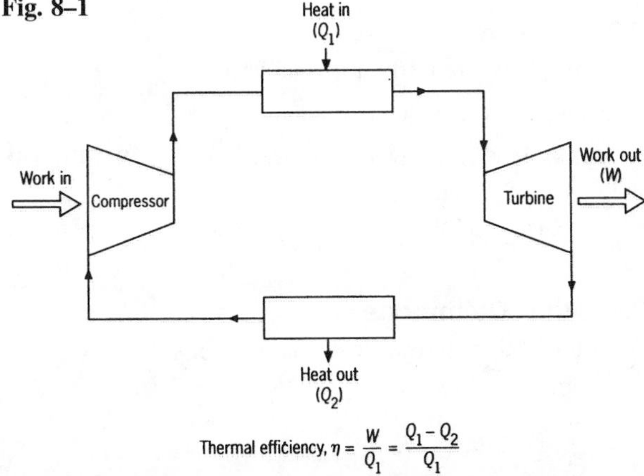

8.2 Entropy

—The existence of entropy follows from the second law.
—Entropy (s) is a property represented by a reversible adiabatic process.
—In the figure (see Fig. 8.2), each p–v line has a single value of entropy (s).

Fig. 8–2

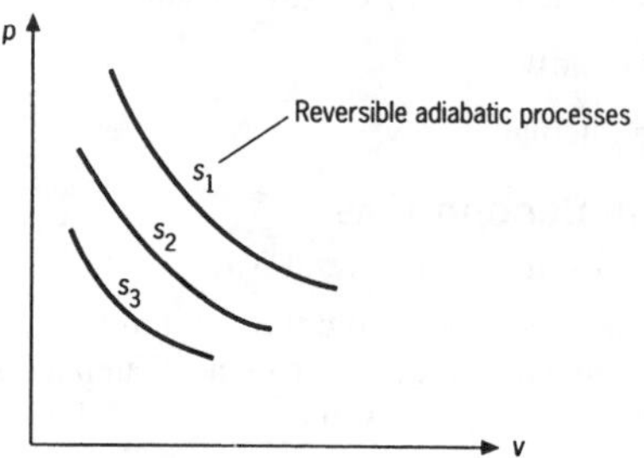

Symbolically, the situation for all working substances is represented by

$$ds = \frac{dQ}{T}$$

where s is entropy.

8.3 Enthalpy

Enthalpy (h) is a property of a fluid itself.

Enthalpy, $h = u + pv$ (units Btu/lbm)

It appears in the steady flow energy equation (SFEE). The SFEE is

$$h_1 + \frac{C_1^2}{2} + Q = h_2 + \frac{C_2^2}{2} + W$$

8.4 Other Definitions

Other useful thermodynamic definitions are:

—A perfect gas follows:

$$\frac{pv}{T} = \text{constant} = R$$

—γ ratio = c_p/c_v (ratio of specific heats) ≅ 1.4
—A constant volume process follows:

$$Q = mc_v\,(T_2 - T_1)$$

—A constant pressure process follows:

$$Q = h_2 - h_1 = mc_p\,(T_2 - T_1)$$

—A polytropic process follows:

$$pv^N = C \text{ and work done} = \frac{p_1v_1 - p_2v_2}{N - 1}$$

8.5 Cycles

Heat engines operate on various adaptations of ideal thermodynamic cycles. These cycles may be expressed on a p–v diagram or T–s diagram, depending on the application (see Fig. 8.3).

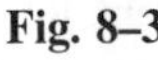
Fig. 8–3

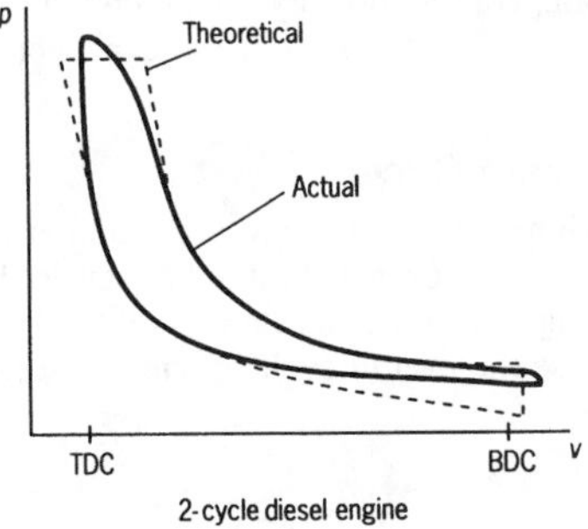

2-cycle diesel engine

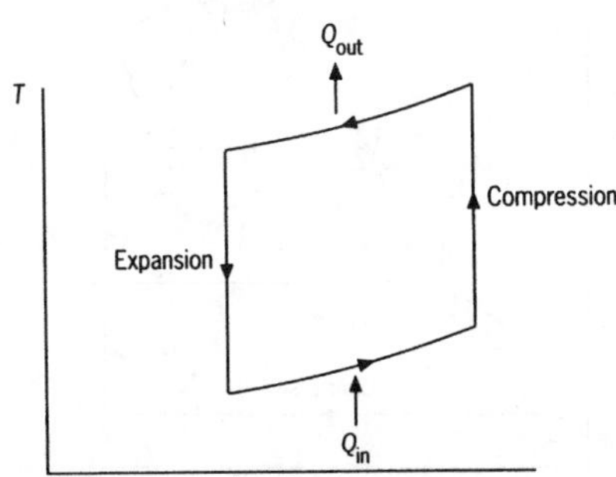

Gas refrigeration cycle

Fig. 8–3 (cont.)

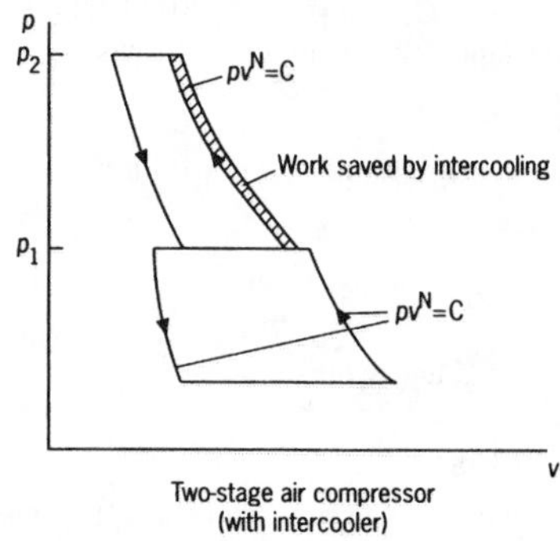

Two-stage air compressor
(with intercooler)

Reciprocating machines such as diesel engines and simple air compressors are traditionally shown on a *p–v* diagram. Refrigeration and steam cycles are better explained by the use of the *T–s* diagram.

8.6 The Steam Cycle

All steam turbine systems for power generation or process use are based on adaptations of the Rankine cycle. Features such as superheating, reheating, and regenerative feed heating are used to increase the overall cycle efficiency (see Fig. 8.4).

Fig. 8–4

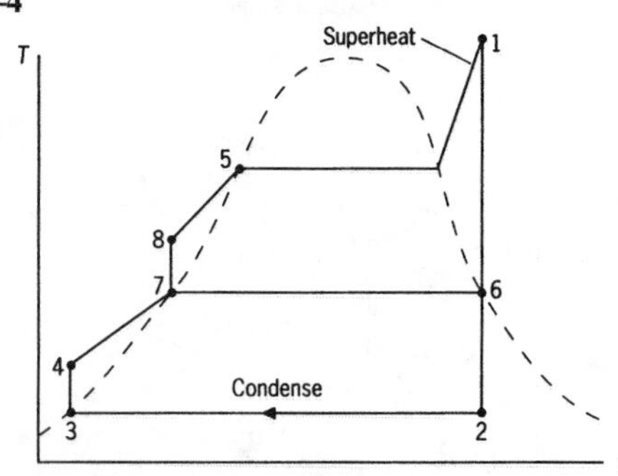

Fig. 8–4 (cont.)

Regenerative steam cycle with superheat and feed heating

8.7 Properties of Steam

Three possible conditions of steam are:

—wet (or "saturated");
—containing a dryness fraction (x);
—superheated ("fully dry").

Standard notations h_f, h_{fg} and h_g are used (see Fig. 8.5).

Published "steam" tables list the properties of steam for various conditions. Two types of table are most commonly used: saturated state properties and superheat properties.

Fig. 8–5

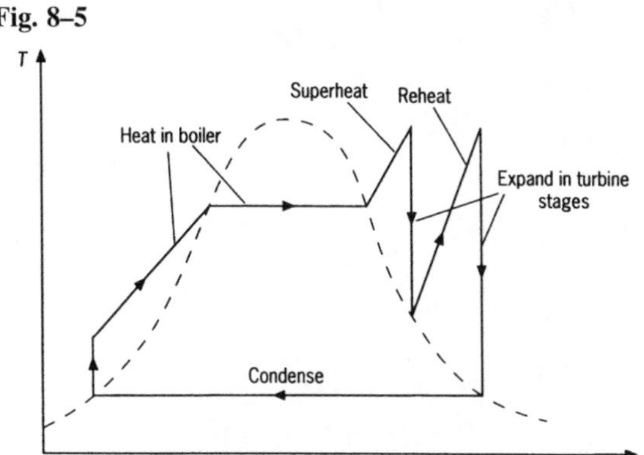

Basic steam cycle with superheat and reheat

Fig. 8–5 (cont.)

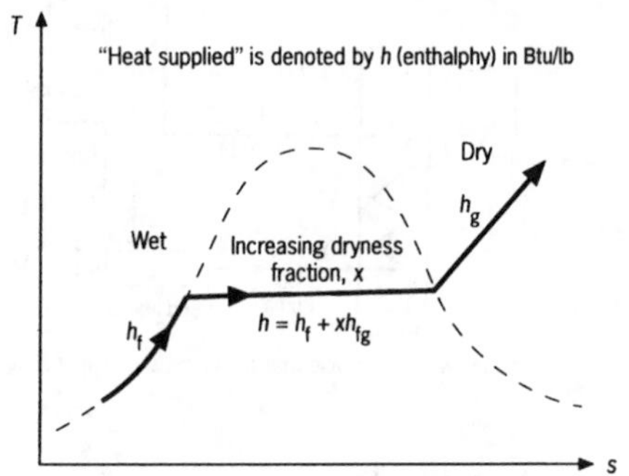

8.7.1 Saturated State Properties

These list the properties corresponding to a range of temperatures (°F) or pressures (lb/in^2) and are formally termed "properties of saturated water and steam."

Note that:

—In the superheat region, pressure and temperature are independent of each other—it is only the t_s that is a function of pressure (see Table 8.2).

Table 8.2

	Abs Pressure	Specific volume	Enthalpy (Btu/lb)			Entropy (Btu/lb.R)		
	P(lb/in^2)	v_g (ft^3/lb)	h_f	h_{fg}	h_g	s_f	s_{fg}	s_g
Example for 212 °F	14.696	26.8	180.15	970.4	1150.5	0.31212	1.4446	1.7567

The equivalent presentation in SI units is:

	Pressure	Saturated Temperature	Specific Volume	Specific enthalpy (kJ/kg)			Specific entropy (kJ/kgK)		
	p(bar)	t_s (°C)	v_g (m^3/kg)	h_f	h_{fg}	h_g	s_f	s_{fg}	s_g
Example for 100 °C	1.01325	100	1.673	419.1	2256.7	2675.8	1.307	6.048	7.355

Note that:

—The maximum pressure listed is 3203.6 psi—known as the *critical pressure*.

—Pressure and temperature are dependent on each other.

Table 8.2 (cont.)

Superheat Properties

These list the properties in the superheat region. The two reference properties are temperature and pressure; all other properties can be derived.

			Temperature (°F)		
			300		*800*
p = 400 lb/in^2	Specific volume v_g = 1.1620 ft^3/lb	v	0.01742	Listed for temperature	1.8163
Sat. temp	Specific enthalpy h_g = 1205.5 Btu/lb	h	270.3	intervals of 100°F	1416.6
t_s = 444.7°F	Specific entropy s_g = 1.4856 Btu/lb.R	s	0.4366	or 200°F	1.6844

Table 8.2 (cont.)

A similar presentation in SI units is:

			Temperature t (°C)		
			300		*600*
$p = 30$ bar	Specific volume $v_g = 0.0666$ m^3/kg	v	0.0812		0.1324
Saturated temperature $t_s = 233.8$ °C	Specific internal energy $u_g = 2603$ kJ/kg	u	2751	Listed for temperature intervals of 50°C or 100°C	3285
	Specific enthalpy $h_g = 2803$ kJ/kg	h	2995		3682
	Specific entropy $S_g = 6.186$ kJ/kgK	s	6.541		7.505

8.8 The Gas Turbine (GT) Cycle

The most basic "open cycle" gas turbine consists of a compressor and turbine on a single shaft. The compression and expansion processes are approximately adiabatic. Figure 8.6 shows the basic (simplified) cycle diagram.

Fig. 8–6

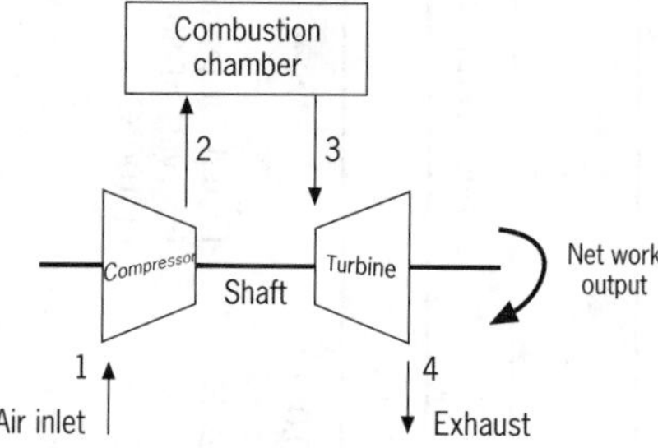

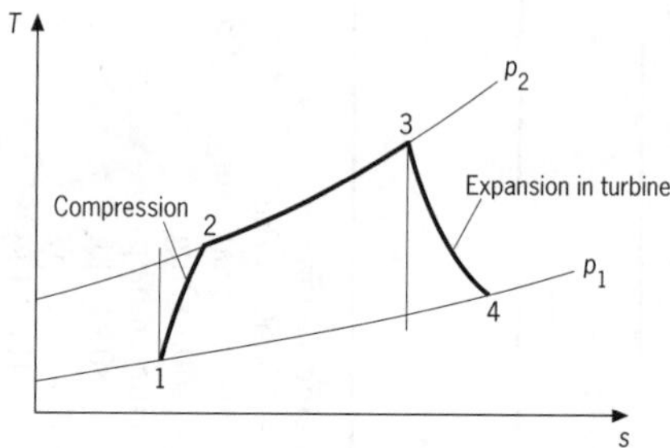

8.9 Useful References

The accepted reference data sources in this field are:

1. Keenan, Keyes, Hilland Moore *Thermodynamic properties of steam*: Pub: John Wiley.
2. Haar. L, Gallagher. J.S. and Kell. G.S. *NBS/NRC Steam Tables: Thermodynamic and Transport Properties and Computer Programs for Vapor and Liquid States* of *Water in S.I Units:* 1984. Pub: Hemisphere Publishing, Washington, D.C.

The European equivalent is:

Rogers and Mayhew: 1994. *Thermodynamic and Transport Properties of Fluids—SI Units.* Pub: Basil Blackwell, UK.

This is a full set of tables, including data on steam, water, air, ammonia, and other relevant fluids.

Web Sites

For descriptions and explanations of thermodynamic cycles: http://fbox.vt.edu:10021/eng/mech/scott/index.html

Section 9

Fluid Mechanics

A fluid is defined as a material that offers no permanent resistance to change of shape. It therefore flows to fill the shape of a vessel in which it is contained. The definition applies to liquids and gases.

9.1 Fluid Pressure

A fluid at rest exerts the same pressure in all directions. The height of a static volume of fluid is called its pressure head or simply its *head* (h) (see Fig. 9.1).

Pressure $= \rho g h$ lbf/in^2 (6894.76 N/m^2)

Fig. 9–1

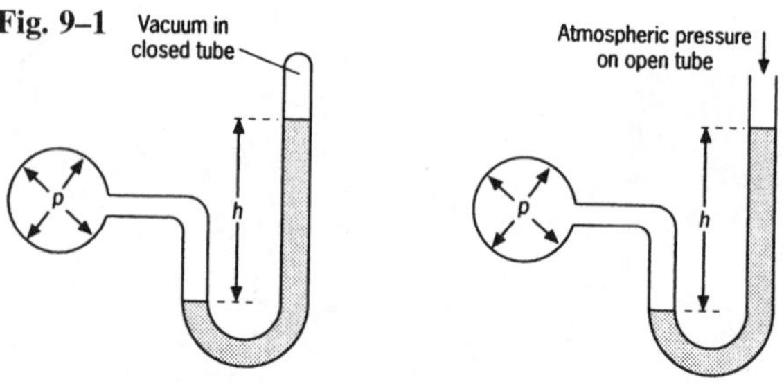

Pressure may be stated as either absolute pressure or gage pressure.

So: p absolute $= p$ gage $+ p$ atmospheric

9.2 Force on Immersed Surfaces

In a tank, the force exerted by a fluid on its sides acts at a point P—known as the *center of pressure* (see Fig. 9.2).

Fig. 9–2

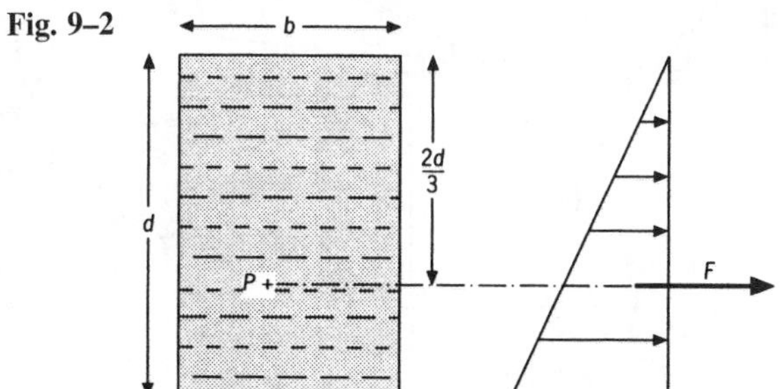

9.3 Bernoulli's Equation

Bernoulli's equation is the result of applying the principle of conservation of energy to fluids.

It is expressed as

Z	+	$v^2/2g$	+	$p/\rho g$	=	constant
↑		↑		↖		↖
Potential head		Velocity head		Static pressure head		Total head

Each of the three terms has the unit of feet—but they are referred to as *heads*, not length. This equation governs the performance of just about all types of fluid machinery and instrumentation.

9.4 The Venturi Meter

A Venturi meter measures the flow in a pipe by sensing the change in pressure caused by a reduction in the cross-section. A U-tube manometer or pressure gage is used to measure the pressure difference (see Fig. 9.3).

Fig. 9–3

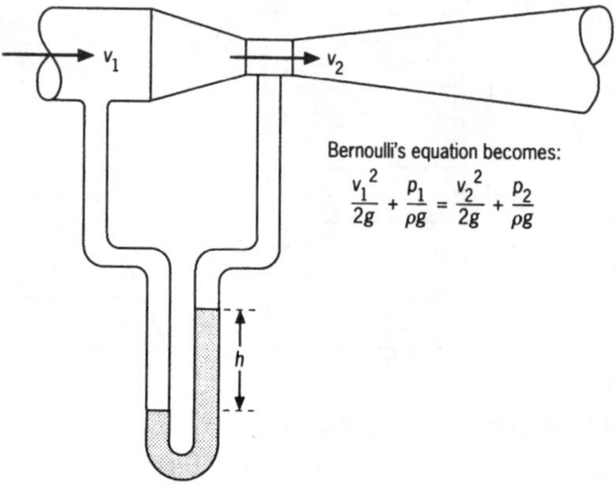

v_1
v_2
h
Bernoulli's equation becomes:
$\frac{v_1^2}{2g} + \frac{p_1}{\rho g} = \frac{v_2^2}{2g} + \frac{p_2}{\rho g}$

9.5 Viscosity

Dynamic viscosity (μ) is a measure of the velocity gradient between stationary and moving parts of a fluid (see Fig. 9.4). It is measured in lbf.s/ft^2 (47.88 Ns/m^2). A further unit of μ is the Poise, i.e., 2.089×10^{-6} lbf.s/ft^2 (0.1 Ns/m^2) or centipoise (cP), which is 0.01 poise.

Fig. 9–4

Velocity (u)
Moving
y
Ratio u/y = dynamic viscosity (μ)
Stationary

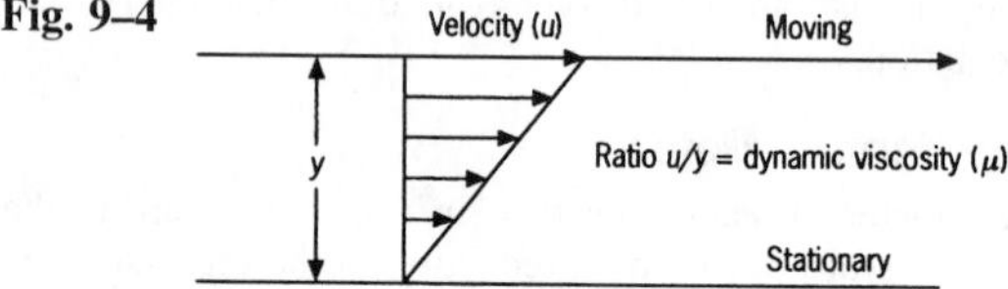

Typical approximate values are shown in Table 9.1.

Table 9.1 Viscosity: Typical Values

Fluid	*lbf.s/ft^2 × 10^{-6}*	*μ(cP)*	*Ns/m^2*
Gear oil	20900	1000	1
Engine oil	2090	100	0.1
Water	20.9	1	0.001
Gasoline	12.5	0.6	0.0006
Air	0.38	0.018	18×10^{-6}

Kinematic viscosity (ν) is a measure of the dynamic viscosity related to density.

Kinematic viscosity, $\nu = \mu/\rho$ ft^2/s (9.290304×10^{-2}m^2/s).

This is commonly used for fuel oils, when units of centistokes, Saybolt seconds universal (SSU), or Engler degrees are used. For lubricating oils it is related to the viscosity grades given in standards such as SAE J300: 1980, ASTM DS 39b, and ISO 3448: 1992.

9.6 Reynolds Number

The expression $\rho v l/\mu$ occurs frequently in fluid mechanics. It is given the name *Reynolds number* (*Re*) and is dimensionless.

$$Re = \frac{\rho vL}{\mu} = \frac{vL}{\nu}$$

ρ = density
v = velocity
ν = kinematic viscosity
L = length

For laminar flow: $Re \leq 2100$
For turbulent flow: $Re \geq 4000$
For $2100 \leq Re \leq 4000$ the flow is *transitional*.

9.7 Aerodynamics

Aerodynamics is a specialism within the general field of fluid mechanics. Much of the subject is related to the performance of airfoil sections (see Fig. 9.5).

Fig. 9–5

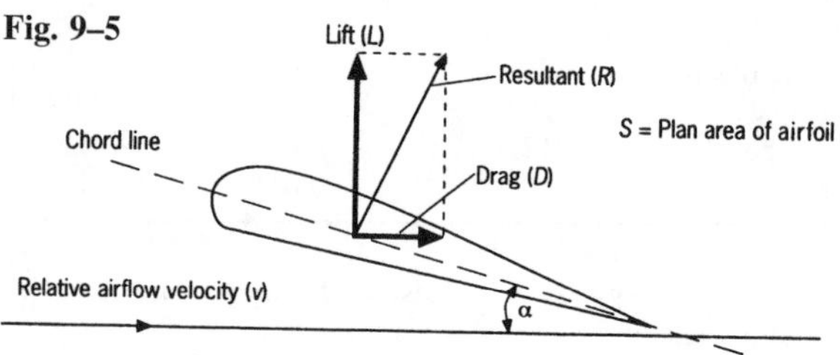

9.8 Lift and Drag Coefficients

Airfoil sections exhibit lift and drag coefficients (see Fig. 9.6).

Lift coefficient $C_L = L/qS$
Drag coefficient $C_D = D/qS$
S = plan area
q = reference pressure ($\frac{1}{2}\,\rho v^2$)

The lift and drag coefficients C_L and C_D vary with the angle of attack (α) of the airfoil. C_L increases to a maximum value and then falls off ("stall"). C_D is small compared to C_L for all angles of attack.

Fig. 9–6

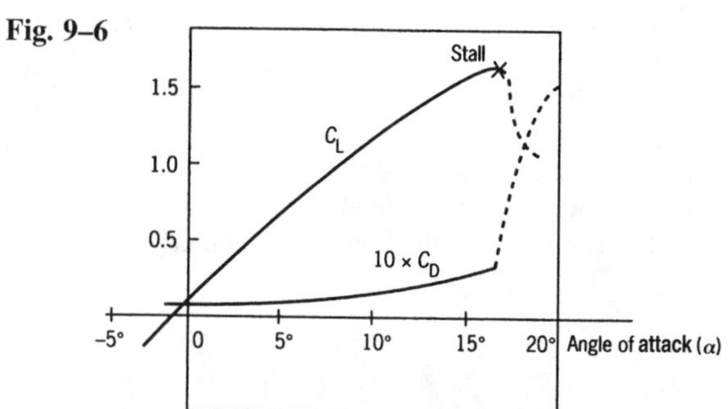

9.9 Mach Number

At high speeds the forces acting on an airplane are affected by compressibility. Under these conditions drag is a function of the ratio between the velocity (V) of the airplane and the velocity of sound (V_s).

i.e., Drag = $\rho V^2 l^2 f(Re, Ma)$

where

$$\text{Mach number } (Ma) = \frac{V}{V_s}$$

9.10 Lubrication

Lubrication in engineering machines can be of several types. The most common type is *hydrodynamic* ("full film") lubrication. The oil film forms a wedge between slightly converging surfaces such as a babbit journal bearing (see Fig. 9.7).

Fig. 9–7

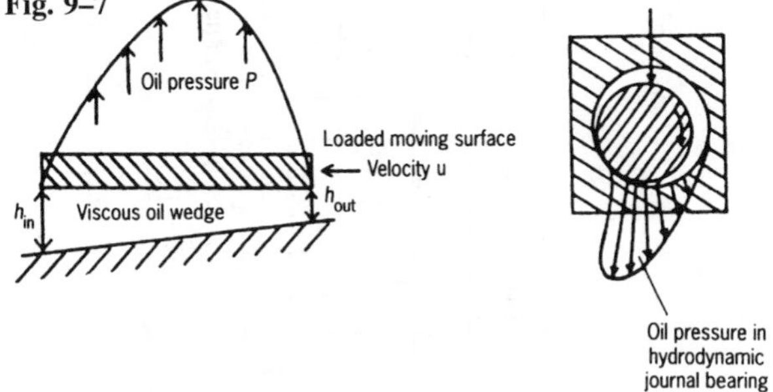

Typical bearing pressures are:

Turbine journals:	580 psi (4 MPa)
Crane hoists:	116psi (0.8 MPa)
Automobile engines:	1450psi (10 MPa)

Items such as pistons and other sliding/reciprocating surfaces often have *boundary* ("mixed") lubrication in which the surfaces are not completely separated.

Hydrostatic ("pressure") lubrication involves the surfaces being separated by a film of fluid when it is under external pressure. A typical application would be the pressure oil system on heavy rotating machines. The study of lubrication is termed "tribology."

9.10.1 Oil Viscosity

Oil viscosity is classified by technical standards such as SAE J300: 1980, or ISO 3448: 1992 (for the "ISO VG" grades) (see Fig. 9.8).

Fig. 9–8

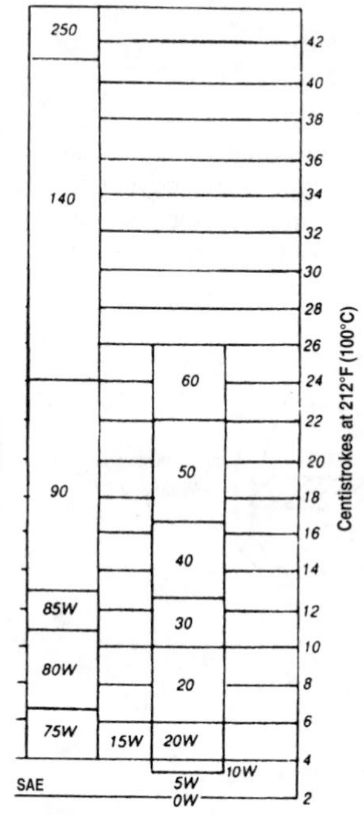

Typical viscosity “grade” data are given in Table 9.2.

Table 9.2 Viscosity “Grade” Data

ISO Viscosity Grade	*Mid-Point Kinematic Viscosity cSt at 104°F (40.0°C)*	*Kinematic Viscosity Limits cSt at 104°F (40.0°C)*	
		Min	**Max**
ISO VG 7	6.8	6.12	7.48
ISO VG 10	10.0	9.00	11.00
ISO VG 15	15.0	13.50	16.50
ISO VG 22	22.0	19.80	24.20
ISO VG 32	32.0	28.80	35.20
ISO VG 46	46.0	41.40	50.60
ISO VG 68	68.0	61.20	74.80
ISO VG 100	100.0	90.00	110.00

9.11 Useful References

ASME standard MFC-3M: 1984: *Measurement of Fluid Flow in Pipes Using Orifices, Nozzles and Venturis.*

Data on fluid flow and pipe friction losses are given in section 3 of *Marke's Standard Handbook for Mechanical Engineers.* Pub: McGraw Hill. ISBN 0-07-004127-X

Web Sites

Society of Automotive Engineers: www.sae.org

Society of Tribologists and Lubrication Engineers: www.stle.org

Section 10

Fluid Equipment

10.1 Turbines

Both steam and gas turbines are in common use for power generation and propulsion. Power ranges are shown in Table 10.1.

Table 10.1 Turbines: Power Ranges

Steam Turbines	*Gas Turbines*
Coal/oil generation: Up to 1,341,000 hp (1000MW)	**Power generation:** Up to 310,000 hp (231 MW)
Nuclear generation: Up to 790,000 hp (589 MW)	**Airplane:** Up to about 40,000 hp (29.8 MW)
Combined cycle application: Up to 41,000 hp (30.6 MW)	**Warships:** Up to about 45,000 hp (33.6 MW)
	Portable power units: Up to about 7,000 hp (5.22 MW)

Both types of turbine are designed by specialist technology licensors and are often built under license by other corporations. Figure 10.1 shows comparative power outputs from other power and transport sources.

Fig. 10–1

COMPARATIVE POWER OUTPUTS

Human 0.067 hp(0.05 kW)

Family auto 55hp(41 kW)

Sports auto 320hp (238.6 kW)

Indy car 750 hp (559.3 kW)

Locomotive 3000 hp (2237.1 kW)

Cargo ship 8000 hp (5965.6 kW)

Warship 70 000 hp (52199 kW)

Light airplane 200 hp(149.1 kW)

Helicopter 550 hp(410.1 kW)

Military airplane 40 000 hp (29828 kW)

Airliner 110 000 hp (82027 kW)

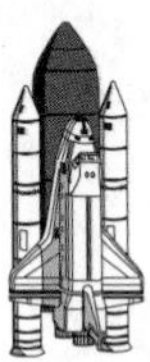

Launch rocket 800 000 hp (596.56 MW)

10.2 Refrigeration Systems

The most common industrial refrigeration plant operates using a vapor compression refrigeration cycle consisting of the standard components of compressor, evaporator, expansion valve, and condenser connected in series (see Fig. 10.2).

Fig. 10–2

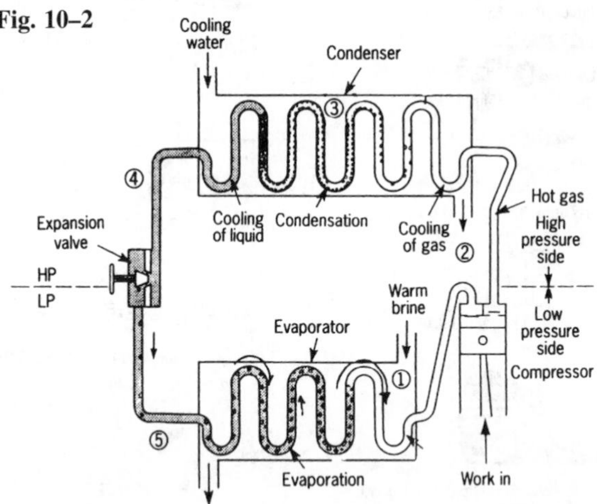

The process can be shown on *T–s* or *P–v* cycle charts (see Fig. 10.3). Performance characteristics are:

Refrigerating effect = RE = $h_1 - h_5$

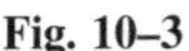
Fig. 10–3

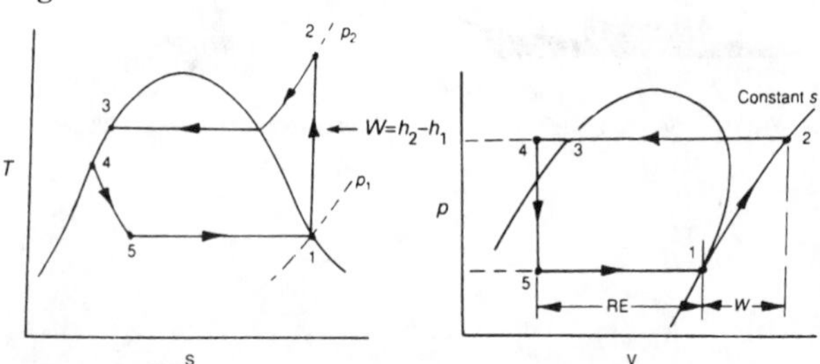

The unit of refrigeration is the "standard ton"

1 Standard ton ≡ Rate of heat absorption of 200 Btu/min (3.5168 kW)

$$- \text{Coefficient of performance (COP)} = \frac{\text{RE}}{W} = \frac{\text{RE}}{h_2 - h_1}$$

Common refrigerants such as R12 (CCl_2F_2) and R22 ($CHClF_2$) still use halogenated hydrocarbons. These are being replaced with other types because of environmental considerations.

10.3 Diesel Engines

10.3.1 Categories

Diesel engines are broadly divided into four categories based on speed (see Table 10.2).

Table 10.2 Diesel Engine Categories

Designation	*Application*	*(Brake) Power Rating*	*RPM*	*Piston Speed ft/s*
Slow speed (2 or 4-cycle)	Power generation, ship propulsion	Up to 70,000 bhp (52.2 MW)	<150	<28
Medium speed (4-cycle)	Power generation, ship propulsion, locomotive propulsion.	Up to 20,000 bhp (14.9 MW)	200–800	<37
High speed stationary engines (4-cycle)	Portable power generation.	Up to 10,000 bhp (7.45 MW)	800–1500	37–53
High speed automobile *(4-cycle)	Trucks, RVs, and some heavy automobiles	Up to 400 bhp (298 kW)	Up to 5000	40–50

*Most U.S. passenger automobiles are gasoline rather than diesel-engine powered. Gasoline engines operate at slightly higher speeds.

10.3.2 Performance

Performance criteria are covered by manufacturers' guarantees. The important ones with typical values are:

Maximum continuous rating (MCR): 100 percent

Specific fuel consumption: 0.36lb/bhp.hr (brake) ($\cong$ 219.5 g/kW.hr)

Lubricating oil consumption: 0.002lb/bhp.hr (brake) ($\cong$ 1.22 g/kW.hr)

NO_x limit: 1400 mg/Nm^3

Note units of specific consumption: lbs/bhp.hr ($\cong$ 609.7 g/kW.hr).

10.4 Heat Exchangers

Heat exchangers can be classified broadly into parallel and counterflow types. Similar equations govern the heat flow. The driving force is the parameter known as logarithmic mean terminal temperature difference $(\Delta t)_{lm}$ °F. Figure 10.4 shows various configurations.

For the parallel flow configuration

$$(\Delta t)_{lm} = \frac{(\Delta t)_1 - (\Delta t)_2}{\ln[(\Delta t)_1/(\Delta t)_2]}$$

where:

A = tube surface area (ft^2)

Δt = temperature difference (°F)

U = overall heat transfer coefficient $\frac{\text{Btu/hr}}{\text{ft}^2.\text{°F}}$

Heat transferred, $q = UA\ (\Delta t)_{lm}$ Btu/hr

For counterflow the same formulae are used. For more complex configurations, such as cross-flow and multi-pass exchangers, $(\Delta t)_{lm}$ is normally determined from empirically derived tables.

Fig. 10–4

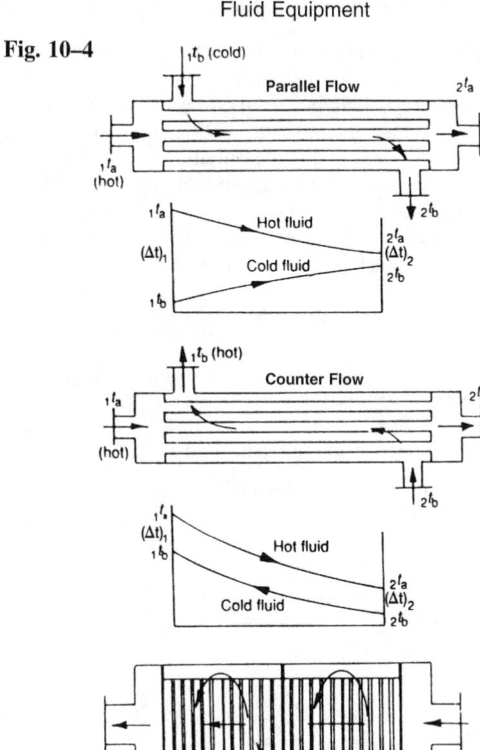

10.5 Centrifugal Pumps

10.5.1 Pump Performance

Pumps are divided into a wide variety of types. The most commonly used are those of the dynamic displacement type. These are mainly centrifugal (radial) but also include mixed flow and axial types. The performance of a pump is mainly to do with its ability to move quantities of fluid. The main parameters are:

- Rotative speed N, rev/min.

- Volume flowrate, Q_v in ft^3/s or gal/min ($1ft^3/s = 449$ gal/min). The SI unit is m^3/h (4.403 gal/min).
- Mass throughput, Q_m in lb/h. The SI unit is kg/s (7936.5 lb/h).
- Head, H in ft. This represents the usable mechanical work transmitted to the fluid and is measured in feet. Together, Q and H define the *duty point* of a pump –a key part of its acceptance guarantee.
- Pump efficiency, η (%) is a measure of the efficiency with which the pump transfers useful work to the fluid.

$$\eta = \text{power output/power input} = \frac{\gamma Q(\text{gal/min})H(\text{ft})}{3960P(\text{bhp})}$$

The equivalent metric formula is: $\eta = \dfrac{\gamma Q(m^3/h)H(m)}{270P(\text{bhp})}$

γ = specific gravity of the pumped fluid

For most centrifugal pumps, the Q/*H* characteristics are as shown in Fig. 10.5.

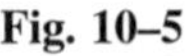

Fig. 10–5

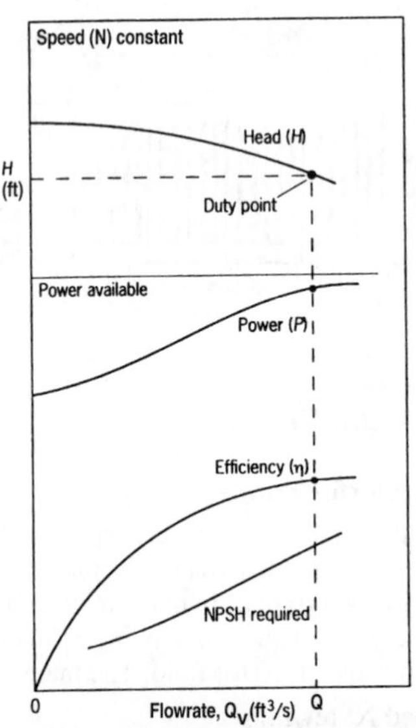

A further performance requirement of a centrifugal pump is its net positive suction head (NPSH), a measure of suction performance at various volume throughputs (see Fig. 10.6).

Fig. 10–6

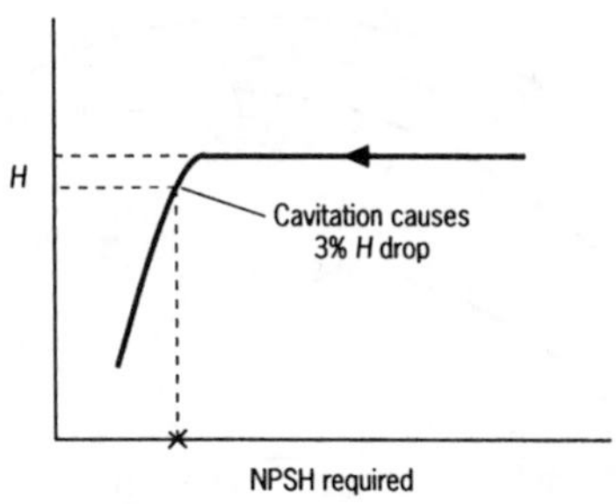

The hydrodynamic performance of centrifugal pumps is covered by the equation:

$$\text{Total head } H = Z_2 - Z_1 + \frac{p_2 - p_1}{\gamma g} + \frac{v_2^2 - v_1^2}{2g}$$

where

Z = distance to a reference plane

γ = density

g = acceleration due to gravity

$$\text{NPSH} = \text{H}_1 + \frac{\text{P atmos}}{\gamma g} - \frac{\text{vapor pressure}}{\gamma g}$$

where

$$\text{H}_1 = \frac{p_1}{\gamma g} + Z_1 + \frac{v_1^2}{2g}$$

10.5.2 Impeller Types

The impeller shape used in a pump is related to the pump's efficiency and a dimensionless "specific speed" (sometimes referred to as "type number") parameter, which is a function of rotative speed, Q_v and H. Figure 10.7 shows approximate design ranges.

Fig. 10–7

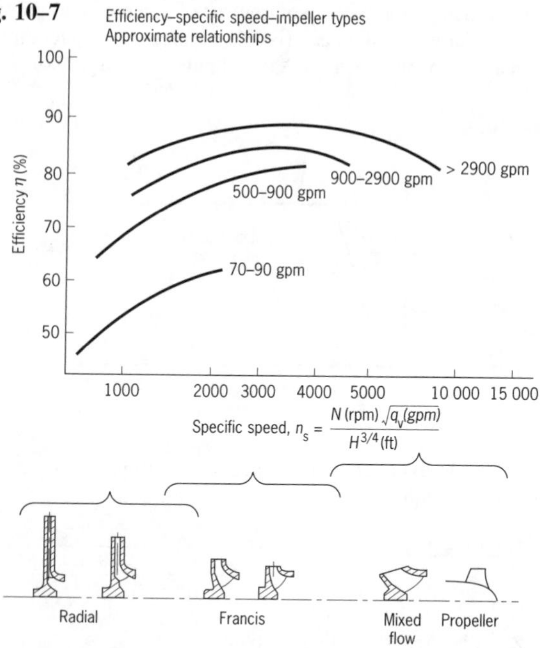
Efficiency–specific speed–impeller types
Approximate relationships
Efficiency η (%)
100
90
80
70
60
50
> 2900 gpm
900–2900 gpm
500–900 gpm
70–90 gpm
1000
2000
3000
4000
5000
10 000
15 000
Specific speed, $n_s = \frac{N\,(\text{rpm})\,\sqrt{q_V(\text{gpm})}}{H^{3/4}\,(\text{ft})}$
Radial
Francis
Mixed flow
Propeller

10.6 Useful References

Standards: Steam Turbines

1. API 611: 1989: *General Purpose Steam Turbines for Refinery Services.*
2. API 612: 1987: *Special Purpose Steam Turbines for Refinery Services.*
3. ANSI/ASME *Performance Test Code No 6:* 1982.
4. EN/ISO 60045-1: 1993. *Guide to Steam Turbine Procurement.*
5. EN/ISO 45510-5-1:1998. *Steam Turbines.*

Standards: Gas Turbines

1. API 616: 1989. *Gas Turbines for Refinery Service.*
2. ANSI/ASME *Performance Test Code 22*: 1985
3. ISO 3977: 1991. *Guide for Gas Turbine Procurement.*

4. ISO 2314: 1989. *Specification for Gas Turbine Acceptance Test.*
5. ISO 11042:1996. *Gas Turbines—Exhaust Gas Emissions*
6. ISO 11086:1996. *Gas Turbines—Vocabulary.*

Gas Turbine Web Sites

International Gas Turbine Institute (IGTI): www.asme.org/igti for links to 20 IGTI committee sites covering all technical aspects of GTs.

Standards: Refrigeration

1. ANSI/ASHRAE standard 15: 1994. *Safety Code for Mechanical Refrigeration.*
2. ANSI/ASHRAE standard 34:1997. *Designation and Safety Classification of Refrigerants.*
3. ISO 917: 1990. *Refrigerant Compressors—Methods of Test for Performance.*

Refrigeration Web Sites

The American Society of Heating, Refrigeration and Air Conditioning Engineers: www.ashrae.org

Air Conditioning and Refrigeration Institute: www.ari.org

Standards: Diesel Engine

1. SAE Standard J3046/1: 1994. *Reciprocating I.C. Engine Performance.*
2. ASTM F1338: 1988. *Standard Guide for Main Propulsion Medium Speed Diesel Engines.*
3. SAE J-2265: 1980. *Diesel Engines Fuel and Performance Requirements—Test Methods for Assessment.*
4. SAE Standard J 1003: 1995. *Diesel Engine Emission Measurement Procedure.*
5. ISO 3046: *Reciprocating Internal Combustion Engines: Performance.*

   It contains the following parts (separate documents):

   ISO 3046/1: *Standard Reference Conditions*
   ISO 3046/2: *Test Methods*
   ISO 3046/3: *Test Measurements*
   ISO 3046/4: *Speed Governing*
   ISO 3046/5: *Torsional Vibrations*
   ISO 3046/6: *Overspeed Protection*
   ISO 3046/7: *Codes for Engine Power*

Engine Web Sites

Society of Automotive Engineers: www.sae.org

Standards: Heat Exchangers

1. TEMA: 1998. *Standards for Design and Construction of Heat Exchangers.* 8TH Edition (Tubular Exchangers Manufacturers Association).
2. API Standard 662 Ed 1: *Plate Heat Exchanger Specifications.*
3. ASME PTC 30: 1998. *Air Cooled Heat Exchangers.*
4. API Standard 661 Ed 4. *Air Cooled Heat Exchangers for General Refinery Service.*
5. API Standard 660 Ed 5. *Shell and Tube Heat Exchangers for General Refinery Service.*
6. JIS (Japan) Standard B8249. *Shell and Tube Heat Exchangers.*

Heat Exchanger Web Sites

Tubular Exchangers Manufacturers Association Inc. (TEMA): www.tema.org

Heat Transfer Research Inc: www.htri-net.com

Standards: Pumps

1. ANSI/HI 1.1-1.2: 2000. *Centrifugal Pumps—Nomenclature, Definitions, Applications and Operation.*
2. ANSI/HI 9.6.1: 1998. *Centrifugal and Vertical Pumps: NPSH Margin.*
3. ASME B73.1 M: 1991. *Horizontal End-Suction Centrifugal Pumps for Chemical Processes.*
4. ISO 2548: 1973. *Specification for Acceptance Tests for Centrifugal Mixed Flow and Axial Pumps—Class C Tests.*
5. ISO 3555: 1977. *Class B Tests.*
6. ISO 5198: 1987. *Precision Class Tests.*

Pump Web Sites

For descriptions of various types of pumps and links to other pump-related sites: www.machinedesign.com/bde/fluid-power/power-input1/fluid.html

Section 11

Pressure Vessels

11.1 Vessel Codes and Standards

Pressure vessels can be divided broadly into "simple" vessels and those that have more complex features. The general arrangement of a simple vessel is as shown in Fig. 11.1—note it has no complicated supports or sections and that the ends are dished, not flat.

Fig. 11–1

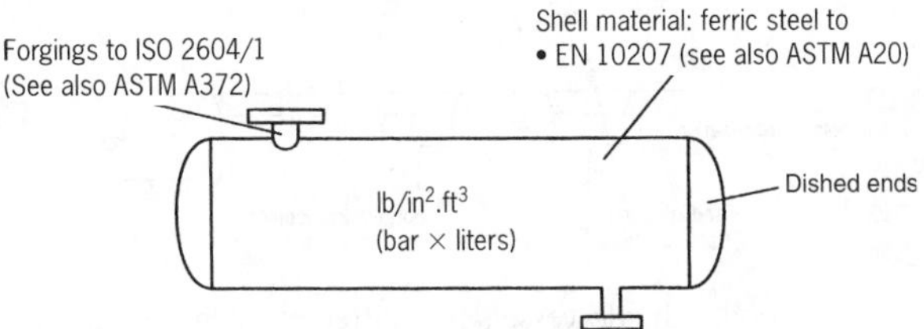

The main international code for simple pressure vessels is: EN 286-1: 1991: *Simple unfired pressure vessels designed to contain air or nitrogen.*

All aspects of designing and manufacturing the vessel are included under the following sections of the standard:

Section 4:	Classification and certification procedures
Section 5:	Materials
Section 6:	Design
Section 7:	Fabrication
Section 8–9:	Welding
Section 10:	Testing
Section 11:	Documentation
Section 12:	Marking

There are three vessel categories, based on capacity in $lb/in^2 \times ft^3$ (bar × liters). More complex pressure vessels follow accepted codes such as: ASME Section VIII: 1998: *Boiler and Pressure Vessel Code*. Fig. 11.2 shows the construction of a typical vessel.

Fig. 11–2

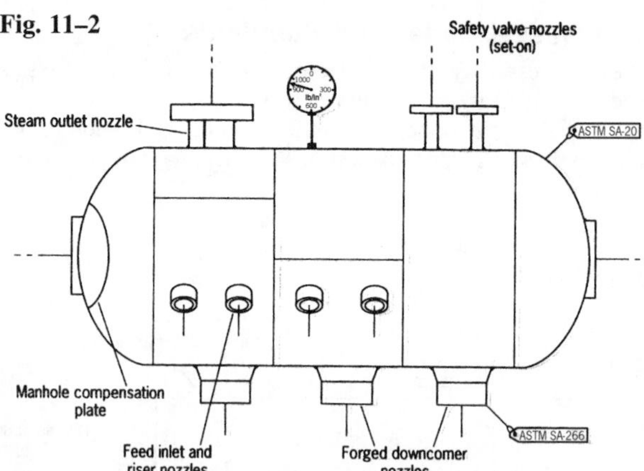

Vessel codes divide vessels into different categories depending on their design, application, and manufacture.

Some key parts of ASME Section VIII: 1998 are shown in Table 11.1.

Table 11.1 Key Parts of ASME Section V111

Subject	*ASME VIII article*
Responsibilities and certification	UG-90
Vessel design features	UW-12, UW-13 and UW-16
Permissible materials of construction	UG-4 (cross-references ASME Section II)
Non-destructive examination (NDE)	UG-93, UW-51
Pressure testing	UG-99
Documentation requirements	UG-115 to UG-120

11.2 Pressure Vessel Design Features

Although straightforward in concept, pressure vessels can exhibit a variety of design features. Different methods of design and assessment are used—all of which are covered in detail in the design codes. Common weld, nozzle, and flange types are as shown in Fig. 11.3.

Fig. 11–3

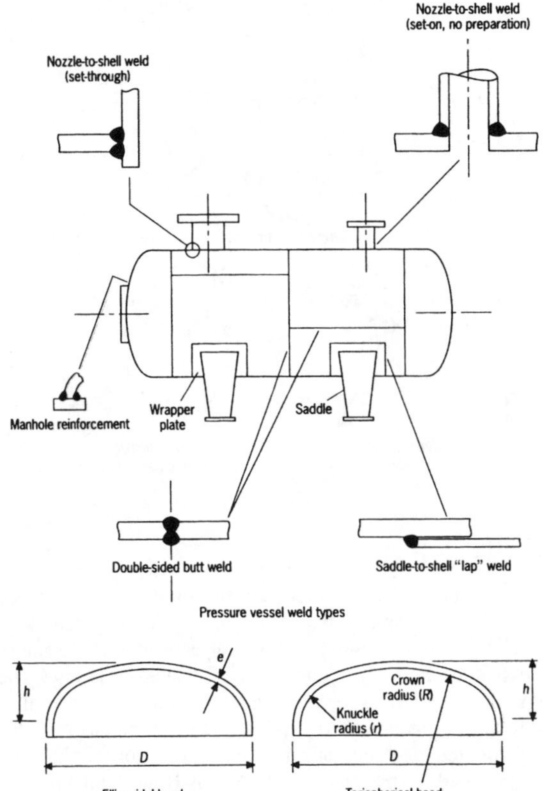

Fig. 11–3 (cont.)

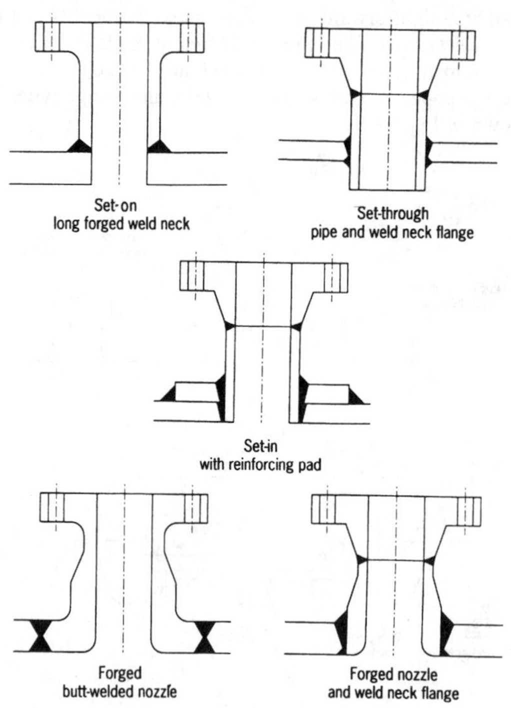

11.3 Vessel Certification

Pressure vessels contain large amounts of stored energy and hence are considered potentially dangerous pieces of equipment—the legislative situation is complex (and varies from state to state), but vessels are normally considered as "statutory items." The ASME Boiler and Pressure Vessel Code contains the basic rules for safe design, manufacture, and operation. It has legal status if fully adopted by a state or other relevant authority. The code is administered by the National Board of Boiler and Pressure Vessel Inspectors (NBBPVI).

11.4 Flanges

Vessel flanges are classified by *type* and *rating*. Flange types are shown in Fig. 11.4.

Fig. 11–4

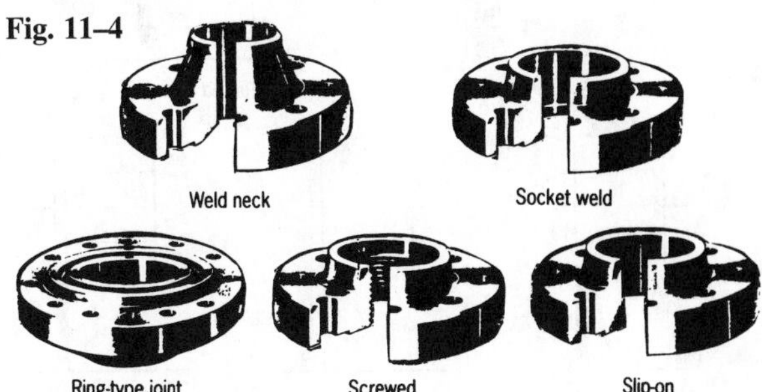

Flanges are rated by pressure (in psi) and temperature, e.g. ANSI B16.5 classes:

150 psi	
300 psi	Detailed size and design
600 psi	information is given in the ANSI
900 psi	B16.5 standard
1500 psi	
2500 psi	

The type of facing is important when designing a flange. Pressure vessel and piping standards place constraints on the designs that are considered acceptable for various applications (see Fig. 11.5).

Fig. 11–5

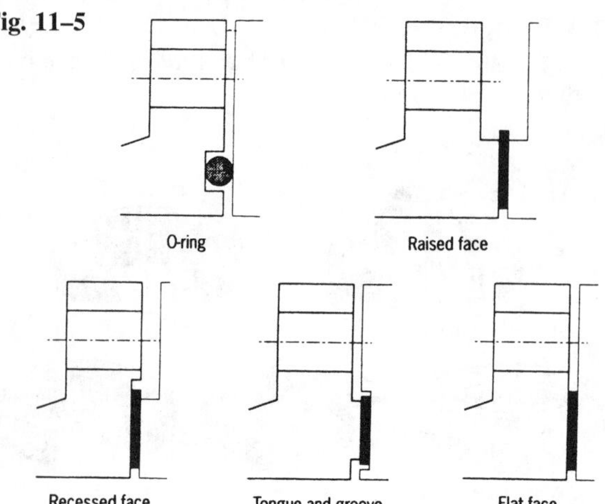

11.5 Useful References

Codes and Standards

1. *The 1998 ASME Boiler and Pressure Vessel Code*: The code contains section I to XI plus related "code case" publications.
2. ASTM A20*: Specification for general requirements for steel plates for pressure vessels.*
3. ASTM E 1419: *Standard test method for examination of seamless, gas-filled pressure vessels.*
4. ASTM A372: *Specification for carbon and alloy steel forgings for thin-walled pressure vessels.*
5. ANSI B 16.24: *Specification for steel flanges.* See also ANSI B 16.5

Web Sites

1. National Board of Boiler and Pressure vessel inspectors: www.nationalboard.org
2. Pressure Vessel Research Council: www.forengineers.org/pvrc/index.html
3. For a list of new European EC directives and standards on pressure equipment, go to www.nssn.org and search using keywords "pressure vessel standards" and "EC."

Section 12

Materials

12.1 Observing Crystals—Order and Disorder

The common idea of a crystal is something that is geometrically regular in shape, is transparent, and has luster. While this is sometimes true, it is not a good general description of a crystalline material.

All metals are crystalline. The basic reason for this is that the molecules are all attracted to each other by "binding forces." These forces are non-directional, giving the tendency to pull the molecules into a regular shape. Every molecule is free to choose where it goes, so it roams around until it finds a location that will make the structure neat and ordered and in which it has the least potential energy. Conceptually, the structure of a metal should be like a neat stack of bricks, rather than a random pile. The neat stack is called a crystal.

All solids have some tendency to be crystalline but some manage it and others don't. Metals form highly regular and packed arrangements of molecules, which can take forms such as body-centered cubic (BCC), face-centered cubic (FCC), and close-packed hexagonal (CPH). Paradoxically, although such crystal structures are an attempt at achieving natural *order*, some metals like to crystallize around an impurity or irregularity of some sort—which you could argue is a search for *disorder*. The existence of dislocations and weakness in materials is proof that a crystal structure, ordered though it is, also contains some disorder at the same time. The science of metallurgy is about trying to improve order (because it makes materials stronger) while also finding, and understanding, the inevitable disorder.

Material properties are of great importance in all aspects of mechanical engineering. It is essential to check the up-to-date version of the relevant US Standards or an international equivalent when choosing or assessing a material. The most common steels in general engineering use are divided into the generic categories of carbon, low-alloy, alloy, and stainless.

12.2 Carbon Steels

The effects of varying the carbon content of plain steels are broadly as shown in Fig. 12.1.

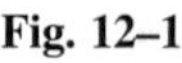

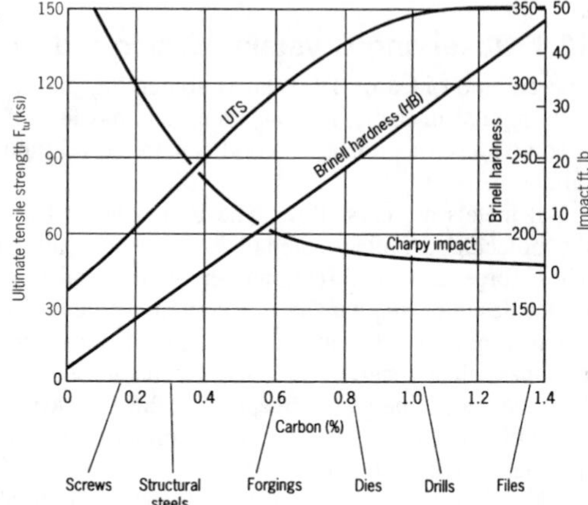

Typical properties are shown in Table 12.1.

Table 12.1 Properties of steels

Type	*%C*	*%Mn*	*Yield F_{ty} (ksi)*	*[R_e*1 MN/m²]*	*Ultimate F_{tu} (ksi)*	*[R_m*2 MN/m²]*
Low C steel	0.1	0.35	32	[220.6]	47	[324]
General structural steel	0.2	1.4	51	[351.6]	75	[517.1]
Steel castings	0.3	—	39	[268.9]	71	[489.5]
Constructional steel for machine parts	0.4	0.75	70	[482.6]	98	[675.7]

*1 R_e (units MN/m² ≡ N/mm² ≡ MPa) is the European designation for yield strength.

*2 R_m (units MN/m² ≡ N/mm² ≡ MPa) is the European designation for ultimate tensile strength.

12.3 Low-Alloy Steels

Low-alloy steels have small amounts of Ni, Cr, Mn, Mo added to improve properties. Typical properties are shown in Table 12.2.

Table 12.2 Low-Alloy Steels

Type	*%C*	*Others (%)*	*Yield F_{ty} (ksi)*	*[R_eMN/m^2]*	*Ultimate F_{tu} (ksi)*	*[R_mMN/m^2]*
Engine crankshafts: Ni/Mn steel	0.4	0.85 Mn 1.00 Ni	70	[482.6]	99	[682.6]
Ni/Cr Steel	0.3	0.5 Mn 2.8 Ni 1.0 Cr	116	[799.8]	132	[910.1]
Gears: Ni/Cr/Mo steel	0.4	0.5 Mn 1.5 Ni 1.1 Cr 0.3 Mo	138	[951.5]	153	[1054.9]

12.4 Alloy Steels

Alloy steels have a larger percentage of alloying elements (and a wider range) to provide strength and hardness properties for special applications. Typical properties are shown in Table 12.3.

Table 12.3 Alloy Steels

Type	*%C*	*Others (%)*	*Yield F_{ty} (ksi)*	*[R_eMN/m^2]*	*Ultimate F_{tu} (ksi)*	*[R_mMN/m^2]*
Chisels, dies C/Cr steel	0.6	0.6 Mn 0.6 Cr	100	[689.5]	126	[868.8]
Heavy-duty dies	2.0	0.3 Mn 12.0 Cr	99	[682.6]	134	[923.9]
Extrusion dies	0.32	1.0 Si 5.0 Cr 1.4 Mo 0.3 V 1.4 W	119	[820.5]	148	[1020.4]
High-speed steel-lathe tools	0.7	4.2 Cr 18.0 W 1.2 V	138	[951.5]	160	[1103.2]
Milling cutters and drills	0.8	4.3 Cr 6.5 W 1.9 V 5.0 Mo	141	[972.2]	175	[1206.6]

12.5 Cast Iron (Ci)

Cast irons are iron/carbon alloys that possess more than about 2% C. They are classified into specific types, as shown in Fig. 12.2.

Fig. 12–2

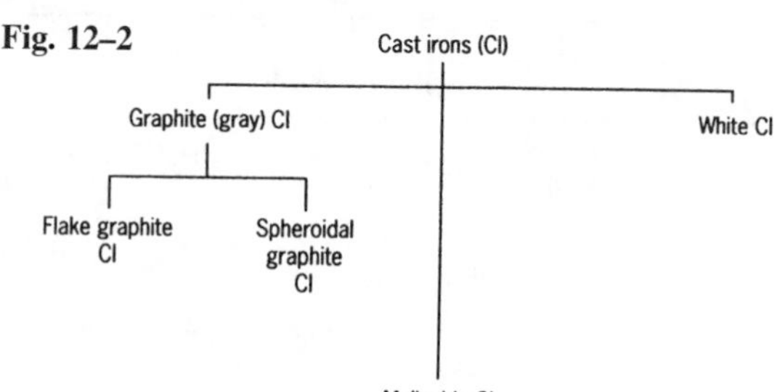

General properties and uses are varied, as shown in Table 12.4.

Table 12.4 Cast Irons

Type	*Ultimate* F_{tu} *(ksi)*	$[R_m(MN/m^2)]$	*Elongation (%)*	*HB*
Grey CI (engine cylinders)	25–53	[172.3–365.4]	0.5–0.8	150–250
Nodular ferritic SG CI (piping)	50–70	[344.7–482.6]	6–16	115–215
Nodular pearlitic SG CI (crankshafts)	87–116	[599.8–799.8]	2–3	210–300
Pearlitic malleable (camshafts and gears)	65–80	[448.2–551.6]	3–8	140–240

12.5.1 Grey CI

These types have a structure of ferrite, pearlite, and graphite, giving a gray appearance on a fractured surface. The graphite can exist as either flakes or spheres. Nodular (SG) CI is obtained by adding magnesium, which encourages the graphite to form into spheres or "nodules."

12.5.2 White CI

This has a structure of cementite and pearlite, making it hard, brittle, and difficult to machine. Its main use is for wear-resisting components. Fracture surfaces have a light-colored appearance.

12.5.3 Malleable CI

These are heat-treated forms of white CI to improve their ductility while maintaining the benefits of high tensile strength.

12.6 Stainless Steels

Stainless steel is a generic term used to describe a family of steel alloys containing more than about 11 percent chromium. The family consists of four main classes, subdivided into about 100 grades and variants. The main classes are austenitic and duplex; the other two—ferritic and martensitic classes—tend to have more specialized application and so are not as commonly found in general use. The basic characteristics of each class are as shown in Table 12.5.

—*Austenitic*: The most commonly used basic grades of stainless steel are usually austenitic. They have 17–25% Cr, combined with 8–20% Ni, Mn, and other trace alloying elements, which encourage the formation of austenite. They have low carbon content, which makes them weldable. They have the highest general corrosion resistance of the family of stainless steels.

—*Ferritic*: Ferritic stainless steels have high chromium content (>17% Cr) coupled with medium carbon, which gives them good corrosion resistance properties rather than high strength. They normally contain some Mo and Si, which encourage the ferrite to form. They are generally non-hardenable.

—*Martensitic*: This is a high-carbon (up to 2% C), low-chromium (12% Cr) variant. The high carbon content can make it difficult to weld.

—*Duplex*: Duplex stainless steels have a structure containing both austenitic and ferritic phases. They can have a tensile strength of up to twice that of straight austenitic stainless steels and they are alloyed with various trace elements to aid corrosion resistance. In general, they are as weldable as austenitic grades but have a maximum temperature limit because of the characteristics of their microstructure.

Table 12.5 Stainless Steels—Basic Data
Stainless steels are commonly referred to by their AISI equivalent classification (where applicable)

AISI	Other classifications	Type*2	Yield F_{ty} (ksi)	[(R_e) MPa]	Ultimate F_{tu} (ksi)	[(R_m) MPa]	E (%) 50 mm	HRB	%C	%Cr	% others*1	Properties
302	ASTM A296 (cast), Wk 1.4300, 18/8, SIS 2331	Austenitic	40	[275.8]	90	[620.6]	55	85	0.15	17–19	8–10 Ni	A general-purpose stainless steel.
304	ASTM A296, Wk 1.4301, 18/8/LC, SIS 2333, 304S18	Austenitic	42	[289.6]	84	[579.2]	55	80	0.08	18–20	8–12 Ni	An economy grade. Not resistant to seawater
304L	ASTM A351, Wk 1.4306 18/8/ELC, SIS 2352, 304S14	Austenitic	39	[268.9]	80	[551.6]	55	79	0.03	18–20	8–12 Ni	Low C to avoid intercrystalline corrosion after welding.
316	ASTM A296, Wk 1.4436 18/8/Mo, SIS 2243, 316S18	Austenitic	42	[289.6]	84	[579.2]	50	79	0.08	16–18	10–14 Ni	Addition of Mo increases corrosion resistance. Better than 304 in seawater.
316L	ASTM A351, Wk 1.4435, 18/8/Mo/ELC, 316S14, SIS 2353	Austenitic	42	[289.6]	81	[558.5]	50	79	0.03	16–18	10–14 Ni	Low C weldable variant of 316.

321	ASTM A240, Wk 1.4541, 18/8/Ti, SIS 2337, 321S18	Austenitic	35	[241.3]	90	[620.6]	45	80	0.08	17–19	9–12 Ni	Variation of 304 with Ti added to improve temperature resistance.
405	ASTM A240/A276/A351, UNS 40500	Ferritic	40	[275.8]	70	[482.7]	30	81	0.08	11 .5–14.5	1 Mn	A general-purpose ferritic stainless steel.
430	ASTM A176/A240/A276, UNS 43000, Wk 1.4016	Ferritic	50	[344.7]	75	[517.1]	30	83	0.12	14 –18	1 Mn	Non-hardening grade with good acid resistance.
403	UNS S40300, ASTM A176/A276	Martensitic	40	[275.8]	75	[517.1]	35	82	0.15	11.5 –13	0.5 Si	Turbine grade of stainless steel.
410	UNS S40300, ASTM A176/A240, Wk 1.4006	Martensitic	40	[275.8]	75	[517.1]	35	82	0.15	11.5 –13.5	4.5–6.5 Ni	Used for machine parts, pump shafts, etc.
—	255 (Ferralium)	Duplex	94	[648.1]	115	[793]	25	280 HV	0.04	24–27	4.5–6.5 Ni	Better resistance to SCC than 316. High strength. Max 575°F (301°C) due to embrittlement
—	Avesta SAF 2507*3, UNS S32750	"Super" Duplex 40% ferrite	99	[682.6]	116	[799.8]	~25	300 HV	0.02	25	7Ni, 4 Mo, 0.3 N	

*1 Main constituents only shown.

*2 All austenitic grades are non-magnetic; ferritic and martensitic grades are magnetic.

*3 Avesta trademark.

12.7 Non-Ferrous Alloys

The term non-ferrous alloys is used for those alloy materials that do not have iron as the base element. The main ones used for mechanical engineering applications, with their ultimate tensile strength ranges, are shown in Table 12.6.

Table 12.6 Property Ranges—Non-Ferrous Alloys

Alloy Type	*Tensile F_{tu} (ksi)*	
Nickel alloys	58–174 ≅	(400–1200 MN/m^2)
Zinc alloys	29–52 ≅	(200–359 MN/m^2)
Copper alloys	29–160 ≅	(200–1103 MN/m^2)
Aluminum alloys	14–72 ≅	(96–496 MN/m^2)
Magnesium alloys	21–49 ≅	(145–338 MN/m^2)
Titanium alloys	58–220 ≅	(400–1517 MN/m^2)

12.8 Nickel Alloys

Nickel is frequently alloyed with copper or chromium and iron to produce a material with high temperature and corrosion resistance. Typical types and properties are shown in Table 12.7.

Table 12.7 Nickel Alloys

Alloy type	*Designation*	*Constituents (%)*	*Ultimate F_{tu} (ksi)*	*[R_m (MN/m^2)]*
Ni-Cu	UNS N04400 ("Monel")	66 Ni, 31 Cu, 1 Fe, 1 Mn	60	[413.7]
Ni-Fe	"Ni lo 36"	36 Ni, 64 Fe	71	[489.5]
Ni-Cr	"Inconel 600"	76 Ni, 15 Cr, 8 Fe	87	[599.9]
Ni-Cr	"Inconel 625"	61 Ni, 21 Cr, 2 Fe, 9 Mo, 3 Nb	116	[799.8]
Ni-Cr	"Hastelloy C276"	57 Ni, 15 Cr, 6 Fe, 1 Co, 16 Mo, 4 W	109	[751.6]
Ni-Cr (age hardenable)	"Nimonic 80A"	76 Ni 20 Cr	116-175	[799.8-1206.6]
Ni-Cr (age hardenable)	"Inco Waspalloy"	58 Ni, 19 Cr, 13 Co, 4 Mo, 3 Ti, 1 AL	116-145	[799.8-999.8]

12.9 Zinc Alloys

The main use for zinc alloys is die casting. The alloys are widely known by their ASTM designations. Typical types and properties are shown in Table 12.8.

Table 12.8 Zinc Alloys

Alloy Type	*Approximate Constituents (%)*	*Ultimate F_{tu} (ksi)*	*[R_m (MN/m^2)]*	*HB*
Alloy "AG40A" (for die casting)	4 Al, 0.03 Mg, 0.25 Cu	35	[241.3]	65
Alloy "AC41A" (for die casting)	4 Al, 0.05 Mg, 1 Cu	40	[275.8]	80

12.10 Copper Alloys

- Copper–zinc alloys are *brasses*
- Copper–tin alloys are *tin bronzes*
- Copper–aluminum alloys are *aluminum bronzes*
- Copper–nickel alloys are *cupronickels*

Perhaps the most common range are the brasses, which are made in several different forms (see Fig. 12.3).

Fig. 12–3

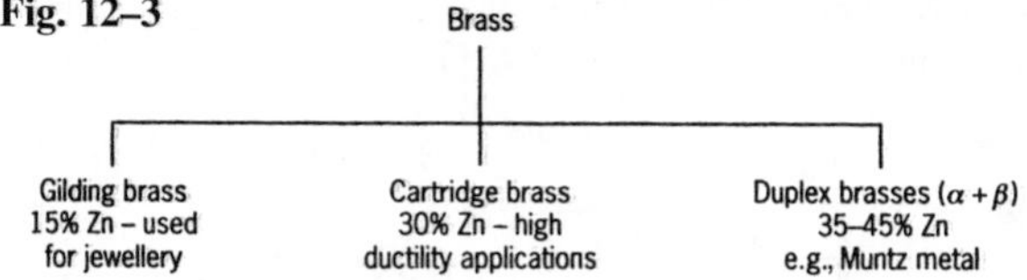

Typical types and properties of copper alloys are shown in Table 12.9.

Table 12.9 Copper Alloys

Alloy type	*Composition (%)*	*Ultimate F_{tu} (ksi)*	*R_m [MN/m^2]*	*HB*
Cartridge brass (shells)	30 Zn	95	[655]	185
Tin bronze	5 Sn, 0.03 P	100	[689.5]	200
Gunmetal (marine components)	10 Sn, 2 Zn	44	[303.4]	80
Aluminum bronze (valves)	5 Al	94	[648.1]	190
Cupronickel (heat exchanger tubes)	10 Ni, 1 Fe	47	[324]	155
Nickel "silver" (springs, cutlery)	21 Zn, 15 Ni	87	[599.9]	180

12.11 Aluminum Alloys

Pure aluminum is too weak to be used for anything other than corrosion-resistant linings. The pressure of relatively small percentages of impurities, however, increases the strength and hardness significantly. The mechanical properties also depend on the amount of working of the material. The basic grouping of aluminum alloys is shown in Fig. 12.4.

Fig. 12–4

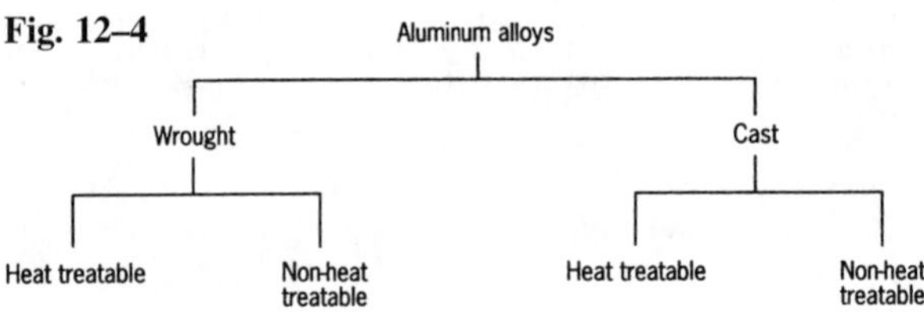

Typical alloy types and properties are shown in Table 12.10.

Table 12.10 Aluminum Alloys

Alloy type (Aluminum Association Designation)	*Constituents (%)*	*Ultimate F_{tu} (ksi)*	*[R_m (MN/m^2)]*	*HB*
5052 (work hardenable)	97.25 Al, 2.5 Mg, 0.25 Cr	28	[193]	47
2011 (heat treatable)	94 Al, 4.5 Cu, 0.5 Pb	55	[379.2]	95
6053 (heat treatable)	97.75 Al, 0.7 Si, 1.3 Mg, 0.25 Cr	16	[110.3]	26

12.12 Titanium Alloys

Titanium can be alloyed with aluminum, copper, manganese, molybdenum, tin, vanadium, or zirconium, producing materials that are light, strong, and have high corrosion resistance. They are all expensive. Typical alloy types and properties are shown in Table 12.11.

Table 12.11 Titanium Alloys

Alloy type	*Constituents (%)*	*Ultimate F_{tu} (ksi)*	*[R_m(MN/m^2)]*	*HB*
Ti-Cu	2.5 Cu	108	[744.7]	360
Ti-Al	5 Al, 2 Sn	128	[882.6]	360
Ti-Sn	11 Sn, 4 Mo, 2 Al, 0.2 Si	190	[1310]	380

12.13 Engineering Plastics

Engineering plastics are widely used in engineering components and are broadly divided into three families: thermoplastics, thermosets, and composites (see Fig. 12.5). Thermoplastic poly-

Fig. 12–5

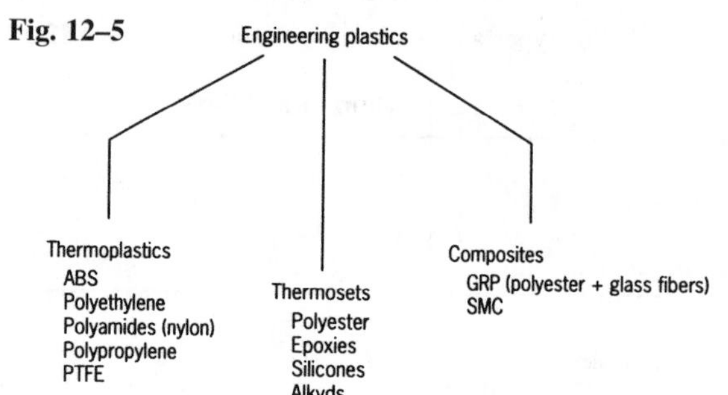

mers can be resoftened by heating, whereas thermosets cannot. Most practical applications of plastic (e.g., automobile body components) need composites to achieve the necessary strength and durability. Typical properties are shown in Table 12.12.

Table 12.12 Engineering Plastics

Type	*Ultimate F_{tu} (ksi)*	*$[R_m$ $(MN/m^2)]$*	*Modulus E (ksi)*	*Modulus $E[(GN/m^2)]$*
PVC	7	[48.3]	508	[3.5]
PTFE	2	[13.8]	44	[0.3]
Nylon	9	[62]	290	[2]
Polyethylene	3	[20.7]	87	[0.6]
GRP	up to 26	[up to 179.3]	up to 2900	[up to 20]
Epoxies	12	[82.7]	1160	[8]

12.14 Material Traceability and Documentation

Material traceability is a key aspect of the manufacture of mechanical engineering equipment. Fabricated components such as pressure vessels are subject to statutory requirements, which include the need for proper material traceability (see Fig. 12.6).

Fig. 12–6

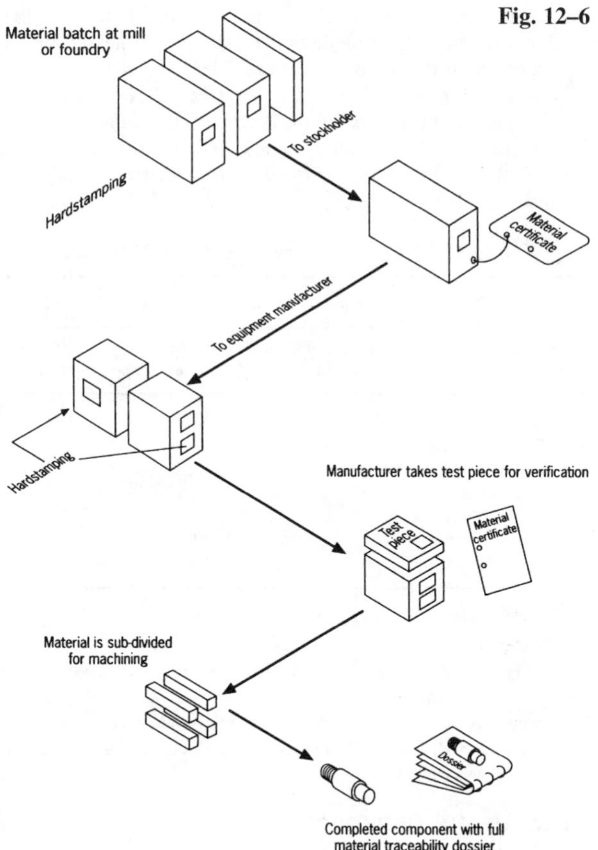

12.14.1 Levels of Traceability: The International Standard EN 10 204

The most common document referenced is the International Standard EN 10 204. It provides for two main "levels" of certification: Class 3 and Class 2. Class 3 certificates are validated by parties other than the manufacturing department of the organization that produced the material—this provides a certain level of assurance that the material complies with the stated properties. The highest level of confidence is provided by the 3.1A certificate, which requires that tests are witnessed by an independent third-party organization. The 3.1B is the most commonly used for "traceable" materials. Class 2 certificates can all be issued and validated by the "involved" manufacturer. The 2.2 certificate is the one most commonly used for "batch" material and has little status above that of a certificate of conformity (see Table 12.13).

Table 12.13 Levels of Traceability : EN 10 204

		Compliance with:			*Test basis*	
EN 10 204 Certificate type	*Document Validation by*	*the order*	*"technical rules"**	*Test results included*	*Specific*	*Non-specific*
3.1A	I	•	•	YES	•	
3.1B	M (Q)	•	•	YES	•	
3.1C	P	•		YES	•	
3.2	P + M (Q)	•		YES	•	
2.3	M			YES	•	
2.2	M			YES		•
2.1	M	•		YES		•

I = an independent (third-party) inspection organization.

P = the purchaser.

M (Q) = an "independent" (normally QA) part of the material manufacturer's organization.

M = an involved part of the material manufacturer's organization.

• Normally the "technical rules" on material properties in the relevant material standard (and any applicable pressure vessel code).

12.15 Useful References

Standards: Carbon Steels

1. ASTM A109/A109M: 1998. *Standard Specification for Cold Rolled Carbon Steel Strips.*
2. ASTM A283/A283M: 1998. *Standard Specification for Low and Intermediate Tensile Strength Carbon Steel Plates.*

Standards: Low-Alloy steels

1. ASTM A514/A514h: 1994. *Standard Specification for High Yield Strength Quenched and Tempered Alloy Steel Plate (Weldable).*

2. Euronorm Standard EN 10083-1: 1991: *Technical Delivery Conditions for Special Steels.*

Standards: Cast Irons

1. ASTM A48: 1994. *Standard Specification for Gray Iron Castings.*
2. ASTM A126: 1995. *Standard Specification of Gray Iron Castings for Valves, Flanges and Pipe Fittings.*
3. ASTM A47/A47M: 1999. *Standard Specification for Ferritic Malleable Iron Castings.*
4. ASTM A220/220M: 1994: *Standard Specification for Pearlitic Malleable Iron.*

Standards: Nickel Alloys

1. ASTM B574:1994. *Specification for Low Carbon Nickel–Chromium And Other Alloys.*
2. ASTM A990: 1998. *Standard Specification for Castings; Iron/Nickel/Chromium and Nickel Alloys.*
3. ASTM B82a: 1999. *Standard Specification for Nickel and Nickel Alloy—Seamless Pipe and Tubes.*
4. SAE J 4070c: 1992. *Wrought Nickel and Nickel-Related Alloys.* This standard contains large numbers of cross-references to other standards.

Standards: Zinc Alloys

1. ASTM B6: 1997. *Specification for Zinc.*
2. ASTM B86: 1996. *Specification for Zinc-Alloy Die Castings.*

Standards: Copper Alloys

1. ASTM B846: 1998. *Standard Terminology for Copper and Copper Alloys.*
2. ASTM B30: 1998. *Standard Specification for Copper-Base Alloys in Ingot Form.*
3. ASTM B36/B36M: 1995. *Standard Specification for Brass Plate, Sheet, Strip and Rolled Bar.*

Standards: Aluminum Alloys

1. ASTM B26/B26M: 1999. *Standard Specification for Aluminum Alloy Sand Castings.*
2. ASTM B108: 1999. *Standard Specification for Aluminum Alloy Permanent Mould Castings.*
3. DIN 1725 (Germany):1998. *Aluminum Casting Alloys.*

Standards: Titanium Alloys

1. MIL-HDBK-5: 1986. *Metallic Materials and Elements for Aerospace Vehicle Structures.*
2. MIL-HDBK-697A NOT 1: 1990. *Titanium and Titanium Alloys.*
3. SAE AMS 238OD: 1994. *Approval and Control of Premium Quality Titanium Alloys.*

Standards: Engineering Plastics

1. MIL-HDBK-700A: 1999. *Plastics.*
2. MIL-I-24 768(2) Sup 1: 1992. *General Specification for Laminated, Thermosetting Plastics.*

Standards: Material Traceability

1. PFI ES41: 1995. *Material Control and Traceability of Piping Components.*
2. ASTM E1338: 1997. *Guide for the Identification of Metals and Alloys in Computerized Material Property Databases.*
3. ASTM E1308: 1992. *Guide for the Identification of Polymers in Computerized Material Property Databases.*

Web Sites

American Iron and Steel Institute (AISI): www.steel.org
Association of Iron and Steel Engineers (AISE): www.aise.org
American Society for Metals International (ASMI): www.asm-intel.org
Society of Plastics Engineers (SPE): www.4spe.org
Plastics Institute of America (PIA): www.eng.uml.edu/Dept/PIA/public_html
Society of the Plastics Industry (SPI): www.plasticsindustry.org

Section 13

Machine Elements

Machine elements is the term given to the set of basic mechanical components that are used as building blocks to make up a mechanical product or system. There are many hundreds of these—the most common ones are shown, subdivided into their common groupings, in Fig. 13.1. The established reference source for the design of machine elements is:

- Shigley. J.E. and Mischke C.R. 1996. *Standard Handbook of Machine Design.* 2nd Ed. Pub: McGraw Hill: ISBN 0-07-056958-4.

13.1 Screw Fasteners

The unified inch, ISO inch, and ISO metric threads are the most commonly used. They are covered by different standards, depending on their size, material, and application.

Figure 13.1 Machine Element Groupings

Locating	*Drives and Mechanisms*	*Energy Transmission*	*Rotary Bearings*	*Dynamic Sealing*
Threaded fasteners	**Shafts**	**Gear trains**	**Rolling**	**Rotating Shaft Seals**
Nuts and bolts	Parallel	Spur	Ball	Face
Set screws	Taper	Helical	Roller (parallel)	Interstitial
Studs	Concentric	Bevel	Roller (tapered)	Axial radial
Grub screws	**Mechanisms**	Worm and wheel	Needle	Bush
Expanding bolts	Crank and sliding	Epicyclic	Self-aligning	Labyrinth
Keys	Ratchet and pawl	**Belt drives**	**Sliding**	Lip ring
Flat	Geneva	Flat	Axial	Split ring
Taper	Scotch-yoke	Vee	Radial	**Reciprocating Shaft Seals**
Woodruff	Carden joint	Wedge	Bush	Piston rings
Profiled	**Cams**	Synchronous	Hydrodynamic	Packing rings
Pins	Constant velocity	**Chain drives**	Hydrostatic	
Split	Uniform acceleration	Roller	Self-lubricating	
Taper	Simple harmonic	Conveyor	Slideways	
Splines	motion (shm)	Leaf		

Figure 13.1 Machine Element Groupings (cont.)

Locating	*Drives and Mechanisms*	*Energy Transmission*	*Rotary Bearings*	*Dynamic Sealing*
Splines	**Clutches**	**Pulleys**		
Retaining rings	Dog	Simple		
Clamps	Cone	Differential		
Clips	Disc	**Springs**		
Circlips	Spring	Tension		
Spring	Magnetic	Compression		
Shoulders and grooves	Fluid coupling			
	Brakes			
	Disk			
	Drum			
	Couplings			
	Rigid			
	Flexible			
	Spring			
	Membrane			
	Cordon			
	Claw			

13.1.1 Unified Inch Screw Threads (ASME B1.1)

Fasteners are defined by their diameter-pitch relationship combination and by their tolerance class, as shown in Table 13.1. Figure 13.2 shows the system of unified inch thread designation (see also Table 13.2 for UNC/UNRC thread dimensions).

Table 13.1 Unified Inch Thread Relationships

Diameter-Pitch Relationship	*Tolerance Class*
UNC and UNRC: Coarse	A represents external threads.
UNF and UNRF: Fine	B represents internal threads.
UN and UNR: Constant pitch	Class 1: Loose tolerances for easy assembly.
UNEF and UNREF: Extra fine	Class 2: Normal tolerances for production items.
	Class 3: Close tolerances for accurate location application.

Fig. 13–2

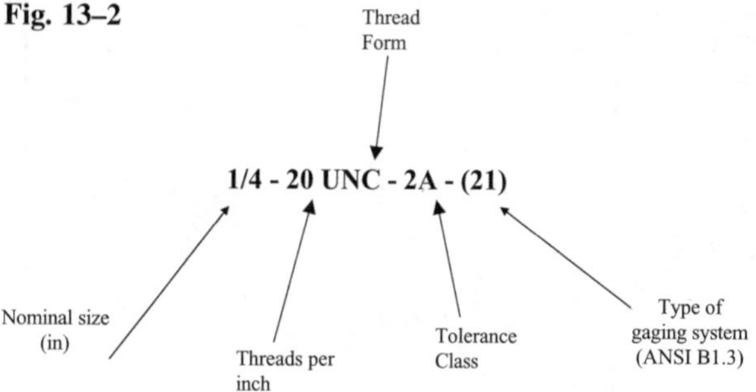

Table 13.2 Typical*UNC/UNRC Thread Dimensions

Nominal size (in)	*Basic major diameter D (in)*	*Threads per inch (in)*	*Basic minor diameter K (in)*
⅛	0.125	40	0.0979
¼	0.25	20	0.1959
½	0.50	13	0.4167
1	1.00	8	0.8647
1½	1.50	6	1.3196
2	2.00	4½	1.7594

**Data from ANSI B1.1: 1982. Equivalent to ISO 5864: 1993*

13.1.2 ISO Metric Screw Threads (ISO 261)

The ISO thread profile is similar to the unified screw thread. They are defined by a set of numbers and letters, as shown in Fig. 13.3(a).

Fig. 13–3a

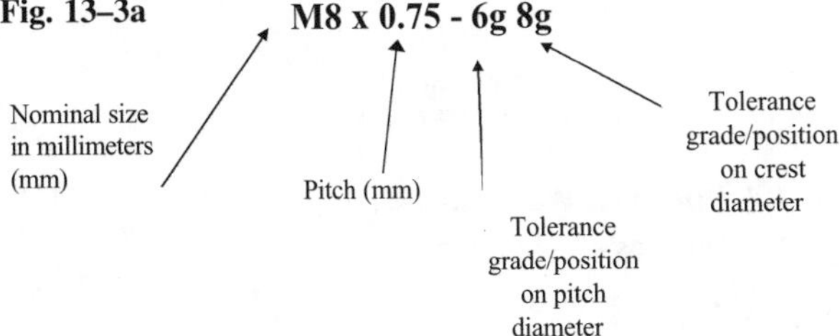

13.1.3 Dimensions: ISO Metric Fasteners (ISO 4759)

Table 13.3/Fig. 13.3(b) show typical dimensions (all in mm) for metric fasteners covered by ISO 4759.

Table 13.3 ISO Metric Fastener Dimensions

		Width A/F (F)		*Head height (H)*		*Nut thickness (m)*	
Size	*Pitch*	*max*	*min*	*max*	*min*	*max*	*min*
M5	0.8	8.00	7.85	3.650	3.350	4.00	3.7
M8	1.25	13.00	12.73	5.650	5.350	6.50	6.14
M10	1.5	17.00	16.73	7.180	6.820	8.00	7.64
M12	1.75	19.00	18.67	8.180	7.820	10.00	9.64
M20	2.5	30.00	29.67	13.215	12.785	16.00	15.57

Fig. 13–3b

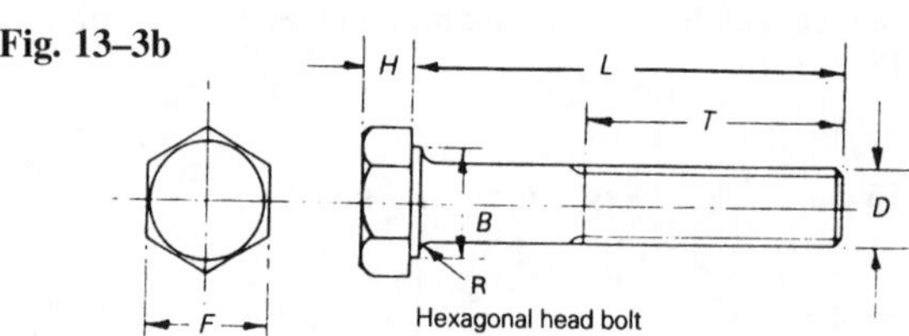

13.1.4 Nuts and Washers

Useful standards are shown in Fig. 13.4.

Fig. 13–4

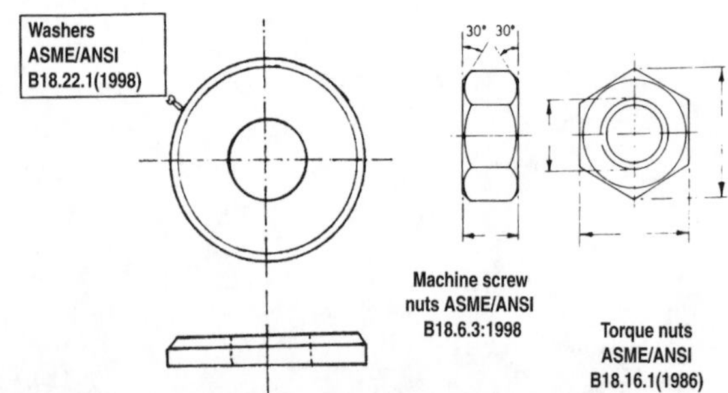

13.2 Bearings

13.2.1 Types

Bearings are basically subdivided into three types: Sliding bearings (plane motion), sliding bearings (rotary motion), and rolling element bearings (see Fig. 13.5). There are three lubrication regimes for sliding bearings:

- Boundary lubrication—there is actual physical contact between the surfaces.
- Mixed-film lubrication—the surfaces are partially separated for intermittent periods.
- Full-film "hydrodynamic" lubrication—the two surfaces "ride" on a wedge of lubricant.

Fig. 13–5

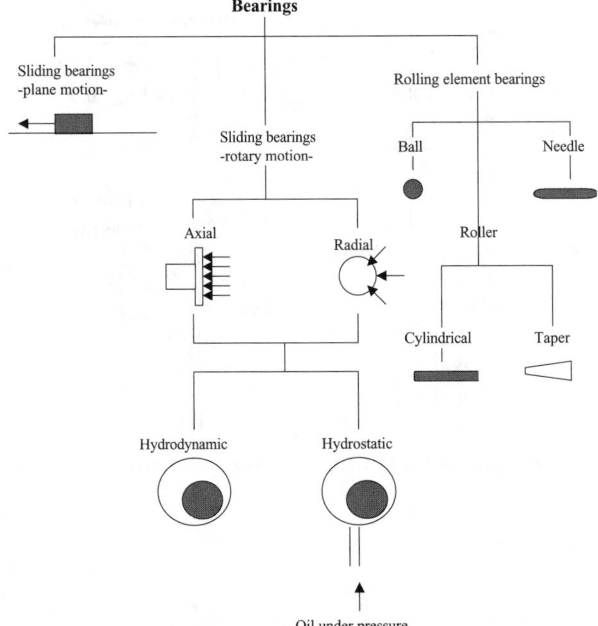

13.2.2 Ball and Roller Bearings

Some of the most common designs of ball and roller bearings are shown in Fig. 13.6. The amount of misalignment that can be tolerated is a critical factor in design selection. Roller bearings have higher basic load ratings than equivalent-size ball types.

Fig. 13–6

Single row radial ball bearing

Allowable misalignment ≃ 0.002 radians

Single row radical roller bearing

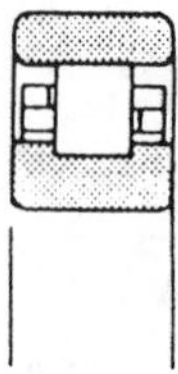

Allowable misalignment ≃ 0.0004 radians

Ball thrust bearing

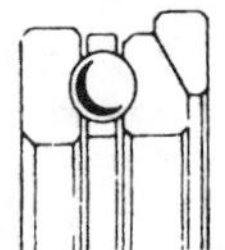

Allowable misalignment ≃ 0.0003 radians

Double row self-aligning ball bearing

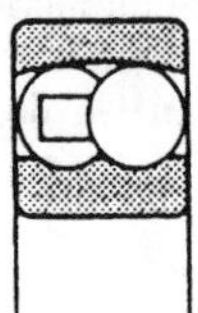

Allowable misalignment ≃ 0.035 radians

Double row spherical roller bearing

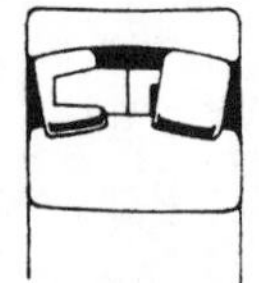

Allowable misalignment ≃ 2°

Tapered roller bearing

Allowable misalignment ≃ 0.0008 radians

13.2.3 Bearing Lifetime

Bearing lifetime ratings are used in purchasers' specifications and manufacturers' catalogs and datasheets. The rating life (L_{10}) is that corresponding to a 10% probability of failure and is given by:

L_{10} radial ball bearings $= (Cr/Pr)^3 \times 10^6$ revolutions

L_{10} radial roller bearings $= (Cr/Pr)^{10/3} \times 10^6$ revolutions

L_{10} thrust ball bearings $= (Ca/Pa)^3 \times 10^6$ revolutions

L_{10} thrust roller bearings $= (Ca/Pa)^{10/3} \times 10^6$ revolutions

Cr and Ca are the static radial and axial load ratings (lb) that the bearing can theoretically endure for 10^6 revolutions. Pr and Pa are corresponding dynamic equivalent radial and axial loads (lb).

So, as a general case:

Roller bearings : L_{10} lifetime $= [16700\ (C/P)^{10/3}]/n$

Ball bearings : L_{10} lifetime $= [16700(C/P)^3]/n$

where

$\left.\begin{array}{l} C = Cr \text{ or } Ca \\ P = Pr \text{ or } Pa \end{array}\right\}$ as appropriate

n = speed in r/min

13.3 Coefficients of Friction

The coefficient of friction between bearing surfaces is an important design criterion for machine elements that have rotating, meshing or mating parts. The coefficient value (f) varies depending on whether the surfaces are static or already sliding, and whether they are dry or lubricated. Table 13.4 shows some typical values.

Table 13.4 Typical Friction (f) Coefficients

	Static (f_o)		Sliding (f)	
Material	*Dry*	*Lubricated*	*Dry*	*Lubricated*
Steel/Steel	0.75	0.15	0.57	0.10
Steel/Cast Iron	0.72	0.20	0.25	0.14
Steel/Phosphor bronze	—	—	0.34	0.18
Steel/Bearing "white metal"	0.45	0.18	0.35	0.15
Steel/Tungsten carbide	0.5	0.09	—	—
Steel/Aluminum	0.6	—	0.49	—
Steel/Teflon	0.04	—	—	0.04
Steel/Plastic	—	—	0.35	0.06
Steel/Brass	0.5	—	0.44	—
Steel/Copper	0.53	—	0.36	0.2
Steel/Fluted rubber	—	—	—	0.05
Cast Iron/Cast Iron	1.10	—	0.15	0.08
Cast Iron/Brass	—	—	0.30	—
Cast Iron/Copper	1.05	—	0.30	—
Cast Iron/Hardwood	—	—	0.5	0.08
Cast Iron/Zinc	0.85	—	0.2	—
Hardwood/Hardwood	0.6	—	0.5	0.17
Tungsten Carbide/Tungsten Carbide	0.2	0.12	—	—
Tungsten Carbide/Steel	0.5	0.09	—	—
Tungsten Carbide/Copper	0.35	—	—	—

Note: The static friction coefficient between similar materials is high, and can result in surface damage or seizure.

13.4 Gear Trains

Gear trains are used to transmit motion between shafts. Gear ratios and speeds are calculated using the principle of relative velocities. The most commonly used arrangements are simple or compound trains of spur or helical gears, epicyclic, and worm and wheel.

13.4.1 Simple Trains

Simple trains have all their teeth on their "outside" diameter (see Fig. 13.7).

Fig. 13–7

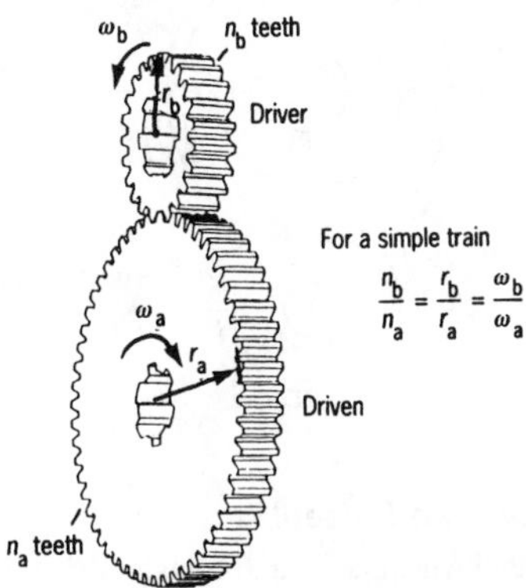

Spur gears – simple train

If an idler gear of radius r_i and n_i teeth is placed in the train, it changes the direction of rotation of the driver or driven gear but does not affect the relative speeds.

13.4.2 Compound Trains

Speeds are calculated as shown in Fig. 13.8.

Fig. 13–8

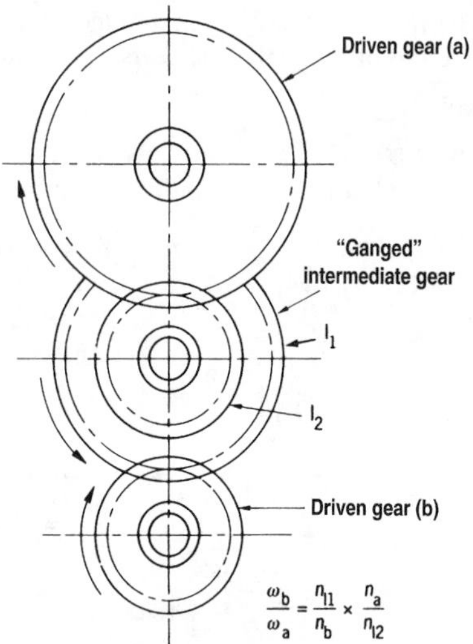

13.4.3 Worm and Wheel

The worm and wheel is used to transfer drive through 90°, usually incorporating a high gear ratio and output torque. The wheel is a helical gear (see Fig. 13.9).

Fig. 13–9

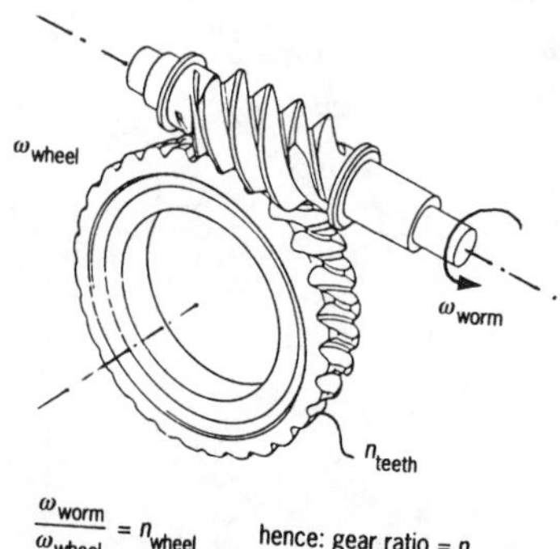

$$\frac{\omega_{worm}}{\omega_{wheel}} = n_{wheel} \qquad \text{hence: gear ratio} = n_{wheel}$$

13.4.4 Double Helical Gears

These are used in most high-speed gearboxes. The double helices produce opposing axial forces, which cancel each other out (see Fig. 13.10).

Fig. 13–10

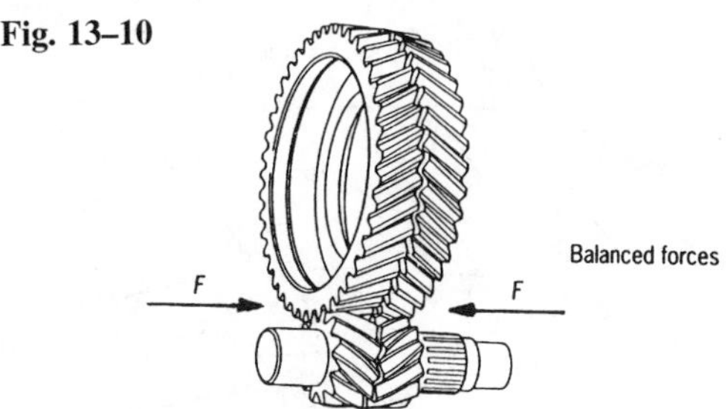

13.4.5 Epicyclic Gears

An epicyclic gear consists of a sun gear on a central shaft, and several planet gears that revolve around it (see Fig. 13.11). A second co-axial shaft carries a ring gear whose internal teeth mesh with the planet gears. Various gear ratios can be obtained depending upon which member is held stationary (by friction brakes). An advantage of epicyclic gears is that their input and output shafts are concentric, hence saving space.

Fig. 13–11

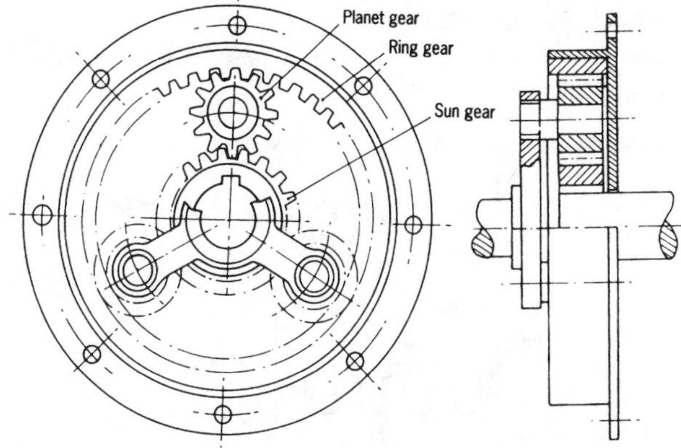

13.4.6 Gear Nomenclature

Gear standards refer to a large number of critical dimensions of the gear teeth. These are controlled by tight manufacturing tolerances (see Fig. 13.12).

Fig. 13–12

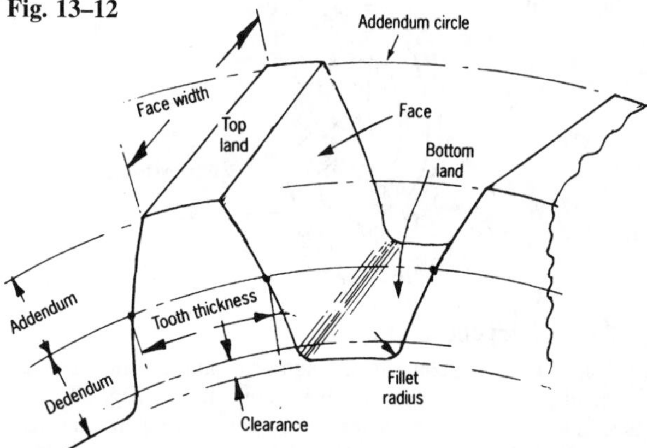

13.5 Seals

Seals are used to seal either between two working fluids or to prevent leakage of a working fluid to the atmosphere past a rotating shaft. They are of several types.

13.5.1 Bellows Seal

This uses flexible bellows to provide pressure and absorb misalignment (see Fig. 13.13).

Fig. 13–13

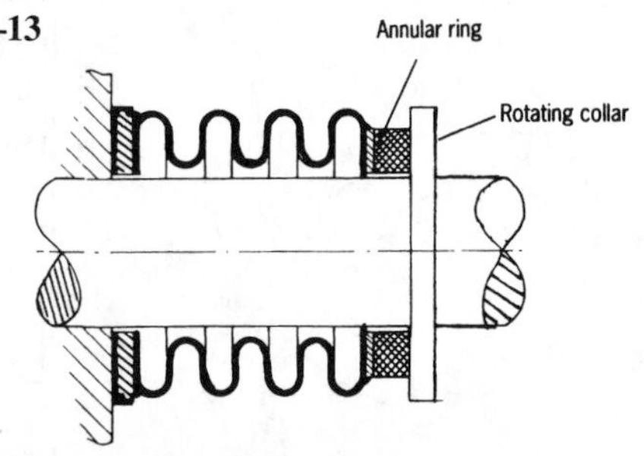

13.5.2 Labyrinth Gland

This consists of a series of restrictions formed by projections on the shaft and/or casing (see Fig. 13.14). The pressure of the steam or gas is broken down by expansion at each restriction. There is no physical contact between the fixed and moving parts.

Fig. 13–14

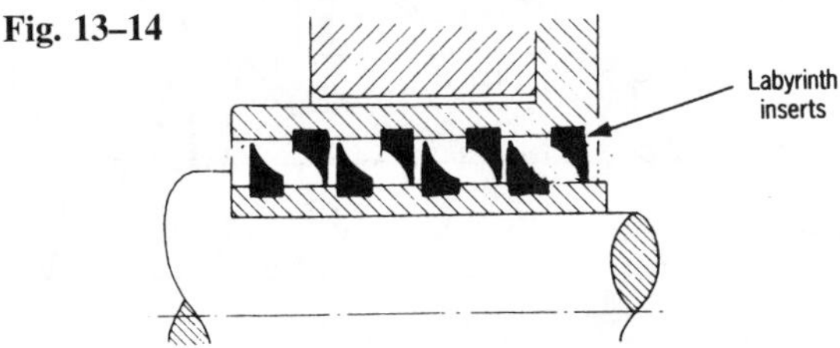

13.6 Mechanical Seals

The key parts of a mechanical seal are a rotating "floating" seal ring and a stationary seat or collar. Both are made of wear-resistant material, and the floating ring is kept under an axial force from a spring (and fluid pressure) to force it into contact with its mating surface.

13.6.1 Mechanical Seal Design Principles

Mechanical seals are used to seal either between two working fluids or to prevent leakage of a working fluid to the atmosphere past a rotating shaft. They can work with a variety of fluids at pressures of up to 7000 lb/in^2 (482.5 MPa) and sliding speeds of more that 800 in/s (20.32 m/s). The core parts of the seal are the rotating "floating" seal ring and the stationary seat (see Fig. 13.15).

Fig. 13–15

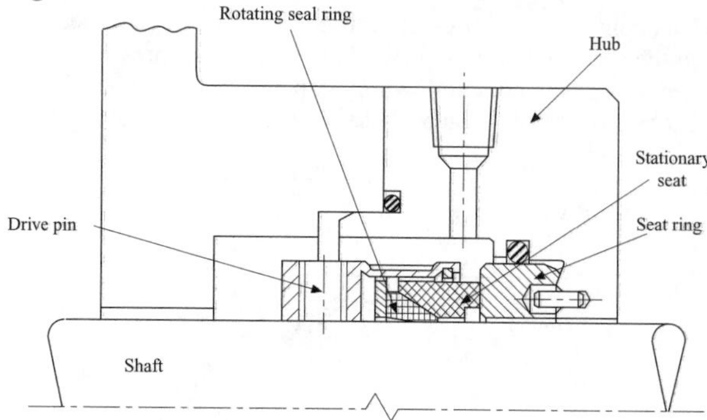

13.7 Shaft Couplings

Shaft couplings are used to transfer drive between two (normally co-axial) shafts. They allow either rigid or slightly flexible coupling, depending on the application.

13.7.1 Bolted Couplings

The flanges are rigidly connected by bolts, allowing no misalignment. Positive location is achieved by using a spigot on the flange face (see Fig. 13.16).

Fig. 13–16

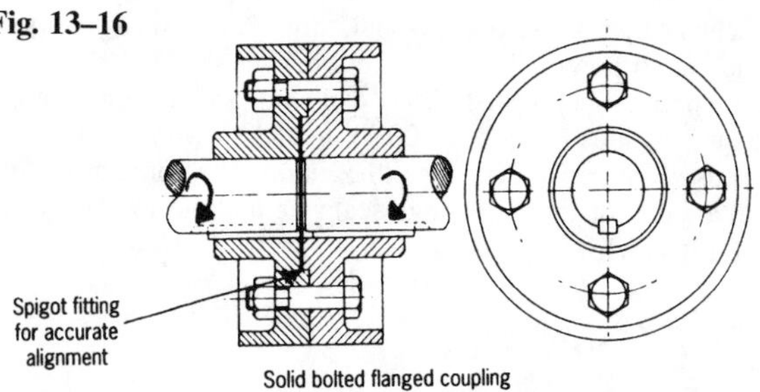

Solid bolted flanged coupling

13.7.2 Bushed-Pin Couplings

Similar to the normal bolted coupling but incorporating rubber bushes in one set of flange holes. This allows a limited amount of angular misalignment (see Fig. 13.17).

Fig. 13–17

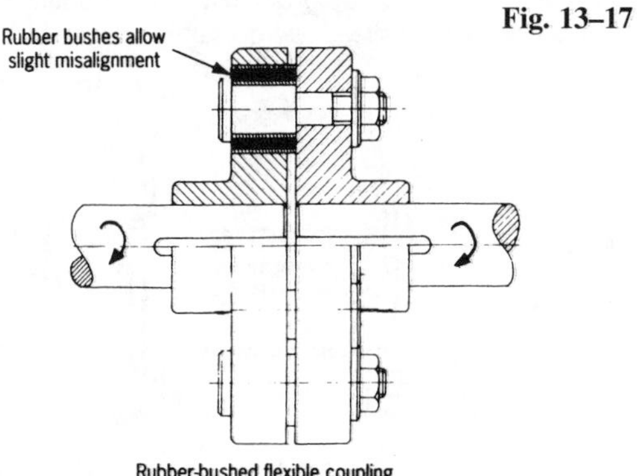

Rubber-bushed flexible coupling

13.7.3 Disc-Type Flexible Coupling

A rubber disc is bonded between thin steel disks held between the flanges (see Fig. 13.18).

Fig. 13–18

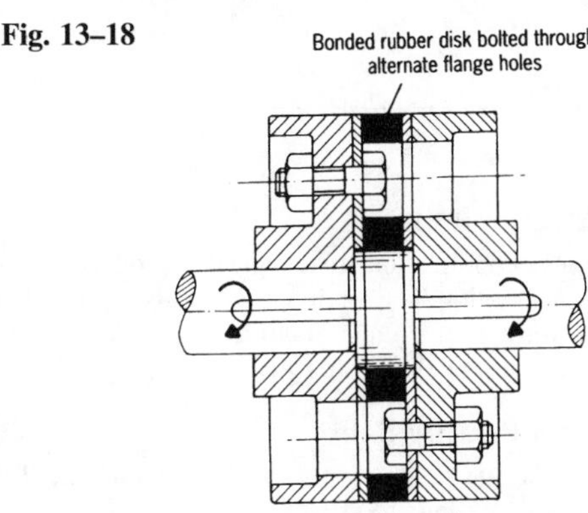

Disk-type flexible coupling

13.7.4 Diaphragm-Type Flexible Couplings

These are used specifically for high-speed drives such as gas turbine gearboxes, turbocompressors, and pumps. Two stacks of flexible steel diaphragms fit between the coupling and its mating input/output flanges (see Fig. 13.19). These couplings are installed with a static prestretch—the resultant axial force varies with rotating speed and operating temperature.

Fig. 13–19

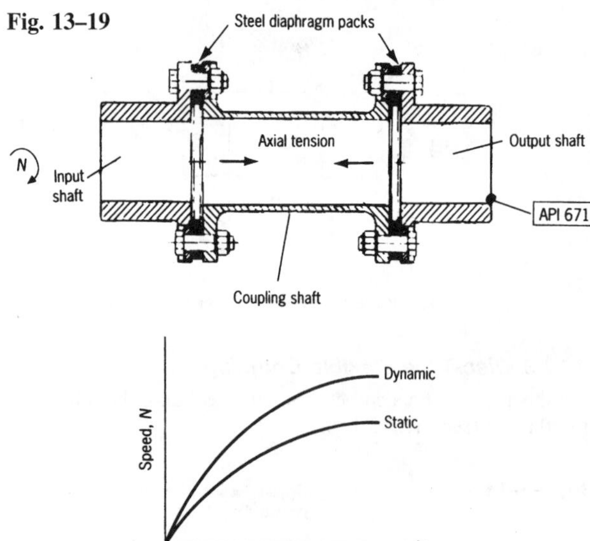

13.8 Cam Mechanisms

A cam and follower combination are designed to produce a specific form of output motion. The motion is generally represented on a displacement/time (or lift/angle) curve. The follower may have knife-edge, roller, or flat profile.

13.8.1 Constant Velocity Cam

This produces a constant follower speed and is suitable only for simple applications (see Fig. 13.20).

Fig. 13–20

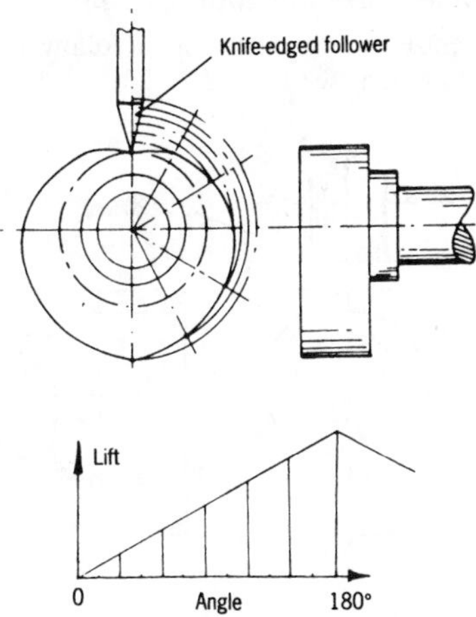

13.8.2 Uniform Acceleration Cam

The displacement curve is second-order function giving a uniformly increasing/decreasing gradient (velocity) and constant d^2x/dt^2 (acceleration) (see Fig. 13.21).

Fig. 13–21

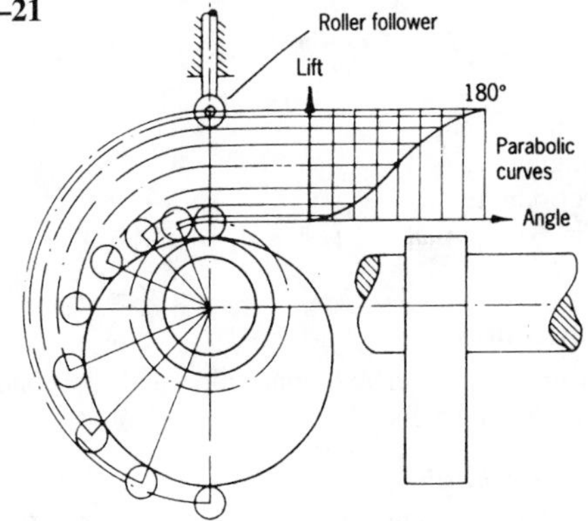

13.8.3 Simple Harmonic Motion Cam

A simple eccentric circle cam with a flat follower produces simple harmonic motion (see Fig. 13.22).

Fig. 13–22

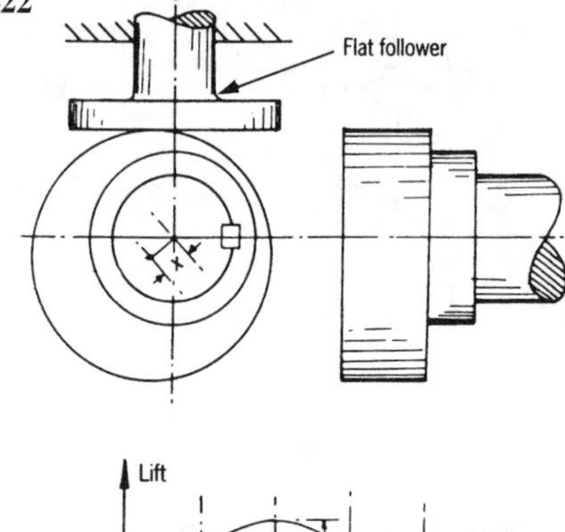

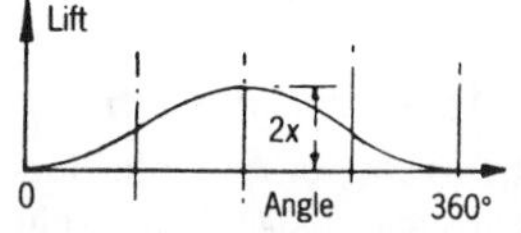

The motion follows the general harmonic motion equations:

$d^2x/dt^2 = -\omega^2 x$

where

x = displacement
ω = angular velocity
T = periodic time
$dx/dt = -\omega a \sin \omega t$
$T = 2\pi/\omega$

13.9 Clutches

Clutches are used to enable connection and disconnection of driver and driven shafts.

13.9.1 Jaw Clutch

One half of the assembly slides on a splined shaft. It is moved by a lever mechanism into mesh, with the fixed half on the other shaft. The clutch can be engaged only when both shafts are stationary. It is used for crude and slow moving machines such as crushers (see Fig. 13.23).

Fig. 13–23

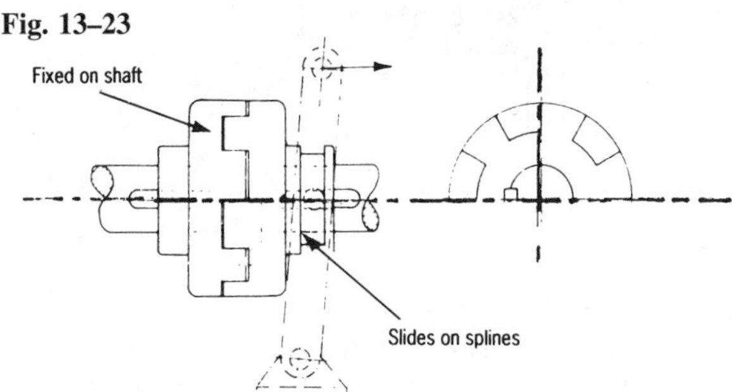

13.9.2 Cone Clutch

The mating surfaces are conical and normally lined with friction material. The clutch can be engaged or disengaged when the shafts are in motion. Used for simple pump drives and heavy-duty materials handling equipment (see Fig. 13.24).

Fig. 13–24

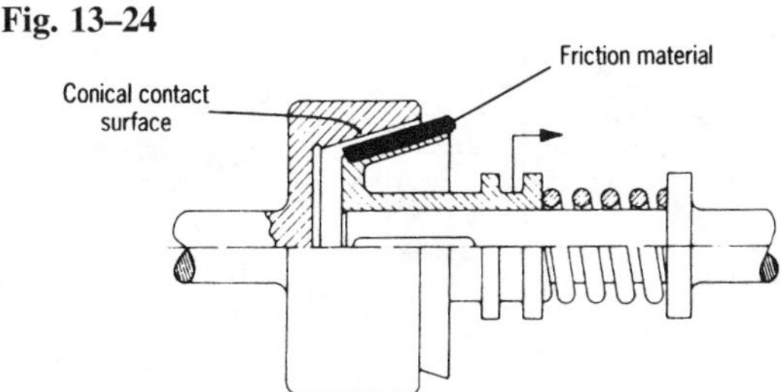

13.9.3 Multi-Plate Disc Clutch

Multiple friction-lined discs are interleaved with steel pressure plates. A lever or hydraulic mechanism compresses the plate stack together. Universal use in automobiles and other motor vehicles with manual transmission (see Fig. 13.25).

Fig. 13–25

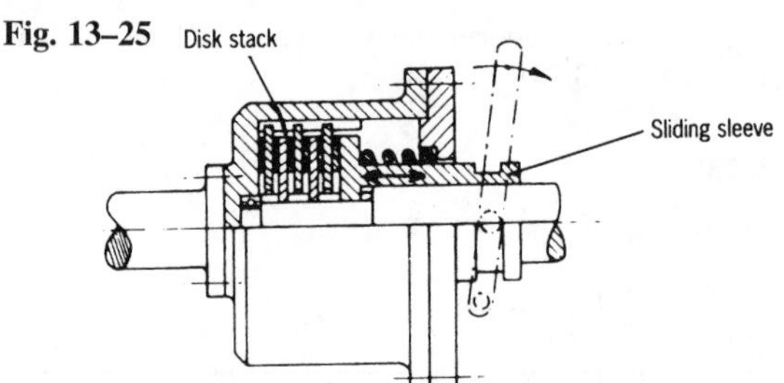

13.9.4 Fluid Couplings

Radial-vaned impellers run in a fluid-filled chamber. The fluid friction transfers the drive between the two impellers. Used in automatic transmission automobiles and for larger equipment such as radial fans and compressors (see Fig. 13.26).

Fig. 13–26

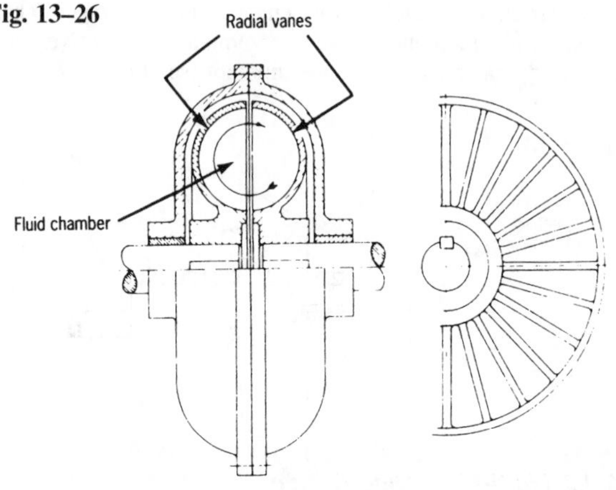

The key design criterion of any type of friction clutch is the axial force required in order to prevent slipping. A general formula is used, based on the assumption of uniform pressure over the contact area (see Fig. 13.27).

Fig. 13–27

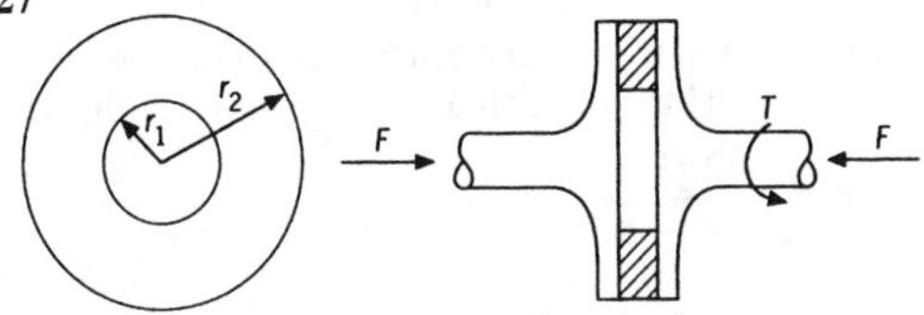

$$\text{Force } F = \frac{3T\ (r_2^2 = r_1^2}{2f(r_2^3 - r_1^3)}$$

T = torque
f = coefficient of friction

13.10 Pulley Mechanisms

Pulley mechanisms can generally be divided into either *simple* or *differential* types.

13.10.1 Simple Pulleys

These have a continuous rope loop wrapped around the pulley sheave. The key design criterion is the velocity ratio (see Fig. 13.28).

Fig. 13–28

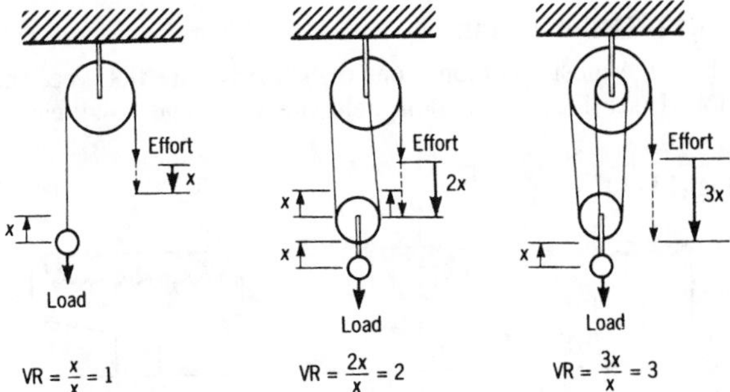

Velocity ratio, VR = the number of rope cross-sections supporting the load.

13.10.2 Differential Pulleys

These are used to lift very heavy loads and consist of twin pulleys "ganged" together on a single shaft (see Fig. 13.29).

$$\mathrm{VR} = \frac{2\pi R}{\pi(R - \mathrm{r})} = \frac{2R}{R - r}$$

Fig. 13–29

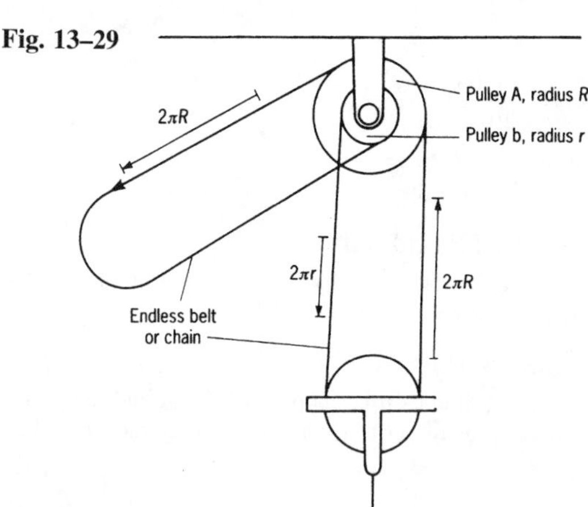

13.11 Drive Types

The three most common types of belt drive are flat, vee, and ribbed (see Fig. 13.30). Flat belts are weak and break easily.

Fig. 13–30

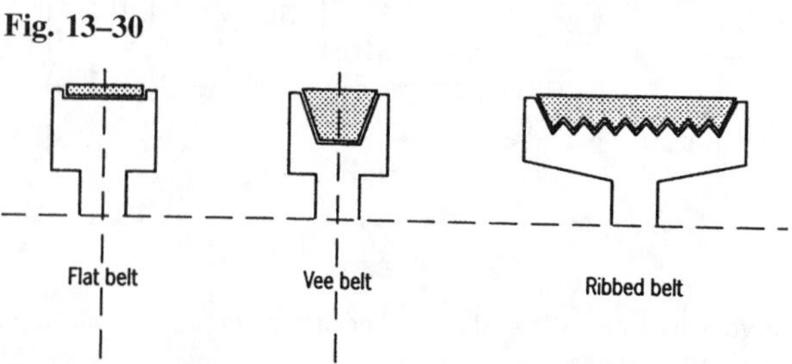

Vee belts can be used in multiples. An alternative for heavy-duty drive is the "ribbed" type, incorporating multiple V-shaped ribs in a wide cross-section.

13.12 Useful References

Standards: Screw Fasteners

1. ANSI B1.1: 1989. *Unified Inch Screw Threads (UN) and UNR Thread Forms.* Equivalent to ISO 5864.
2. ANSI B1.3: 1988: *Screw Thread Gaging Systems for Dimensional Acceptability.*
3. ISO 261: 1998: *ISO General Purpose Metric Screw Threads—General Plan.*
   Part 1: *Principles and Basic Data.* (Gives data for diameter from 1.0–300 mm; see also ISO 68/ISO 261/ISO 965.)
   Part 2: *Specification for selected limits of size.* (Gives size data for ISO coarse threads diameter 1.0–68 mm and ISO fine threads diameter 1.0–33 mm.)
4. ISO 4759-1: 1985. *Tolerances for Fasteners.*
5. ISO 4759.3: 1984: *Washers for Metric Bolts.*

Blake. A. *What Every Engineer Should Know About Threaded Fastener Materials and Design*: 1986: Pub: Marcel Dekker Inc.

Standards: Bearings

1. ABMA A24.2: 1995. *Bearings of Ball, Thrust and Cylindrical Roller Types—Inch Design.*
2. ABMA A20: 1985: *Bearings of Ball, Radial, Cylindrical Roller, and Spherical Roller Types—Metric Design.*
3. ISO 8443: 1989: *Rolling Bearings.*
4. ANSI/ABMA/ISO 5597: 1997: *Rolling Bearings—Vocabulary.*

Bearings Web Sites

Anti-Friction Bearing Manufacturers Association Inc: www.afbma.org

www.skf.se/products/index.htm

www.nsk-ltd.co.jp

Standards: Gears

1. ISO 1328: 1975. *Parallel Involute Gears—ISO System of Accuracy.*
2. ANSI/AGMA 2000-A88. 1994. *Gear Classification and Inspection Handbook.*

3. ANSI/AGMA 6002: B93. 1999. *Design Guide for Vehicle Spur and Helical Gears.*
4. ANSI/AGMA 6019-E89. *Gear-Motors Using Spur, Helical Herringbone Straight Bevel and Spiral Bevel Gears.*

Gear Web Sites

www.agma.org
www.Reliance.co.uk
www.flender.com

Standards: Seals

Mechanical seals are complex items and manufacturers' in-house (confidential) standards tend to predominate. Some useful related standards are:

1. MIL S-52506D. *Mechanical Seals for General Purpose Use.*
2. KS B1566: 1997. *Mechanical Seals.*
3. JIS (Japan) B2405: 1991. *Mechanical Seals—General Requirements.*

Seal Web Sites

www.flexibox.com
www.garlock-inc.com

Standards: Couplings

1. API 617: 1990. *Special Purpose Couplings for Refining Service.*
2. AGMA 515. *Balance Classification for Flexible Couplings.*
3. KS B1555. *Rubber Shaft Couplings.*
4. KS B1553: *Gear Type Shaft Couplings.*
5. ISO 10441: 1999. *Flexible Couplings for Mechanical Power Transmission.*
6. ANSI/AGMA 9003-A91 (R1999). *Flexible Couplings—Keyless Fits.*

Standards: Clutches

1. SAE J2408: 1984: *Clutch Requirements for Truck and Bus Engines.*

Standards: Pulleys

1. SAE - J636: 1992. *V-Belts and Pulleys.*

Standards: Belt Drives

1. SAE J637: 1998. *V-Belt Drives.*
2. RMA-IP20: 1987. *Specification for Drives—V Belt and Sheaves.*
3. ISO 22:1991. *Belt Drives–Flat Transmission Belts.*
4. ISO 4184:1992. *Belt Drives—Classical and Narrow V-Belts.*

Standards: Wire Ropes

1. ASTM A93i: 1996. *Test Methods for Wire Ropes.*
2. SIS (Sweden) 765 30 01: 1984. *Steel Wire Ropes for Cranes.*
3. ISO 2408: 1985. *Steel Wire Ropes for General Purposes.*

See also: *Wire Rope Users Manual* published by AISI.

Standards: Springs

1. ASTM A407: 1998. *Standard Specification for Steel Wire Coiled Springs.*
2. SAE J1122: 1994. *Helical Springs Specification Check Lists.*

Standards: Pins

1. ASME B18.8.2: 1995. *Taper Pins, Dowel Pins, Straight Pins, Grooved Pins, and Spring Pins (Inch Series).*
2. ISO 2339: 1986. *Taper Pins—Unhardened.*

Standards: Splines

1. ANSI B92.2M: 1989. *Metric Involute Splines.*
2. ASME B18.25.1M: 1996. *Square and Rectangular Keys and Keyways.*

Standards: Woodruff Keys

1. ASME B18.25M: 1996. *Woodruff Keys and Keyways.*

Section 14

Design and Production Tools

14.1 Quality Assurance: QS-9000 and ANSI/ ISO/ASQC Q9000: 1994 ("ISO 9000")

Quality standards have been around for many years. Their modern day development started with the U.S. Military and major automobile manufacturers. The American Society for Quality (ASQ) has played a key role working with the International Organization for Standardization (ISO) and CEN (the national standards organization of the European countries) in developing ISO 9000 series standards as an effective model for quality assurance. With recent harmonization of standards under the EN classification, these standards have obtained (almost) universal status as *the* standard of quality assurance.

14.1.1 What Is QS-9000?

QS-9000 is the common supplier quality standard adopted by the three main U.S. automobile manufacturers. It is based on ISO 9001: 1994 but contains specific interpretations relating to suppliers of selected auto parts and services.

14.1.2 What Is ANSI/ISO/ASQC-Q9000:1994? ("ISO 9000")

The ISO 9000 documents were designed with the principle of "universal acceptance" in mind, with the capability of being developed in a flexible way to suit the needs of many different businesses and industries. The series was developed in the form of a number of sections, termed "models"—intended to fit in broadly with the way that industry is structured. Inevitably, this has resulted in a high level of *generalization*. The current "models" are as follows:

1. ANSI/ISO/ASQC Q9001:1994 (ISO 9001) *Quality Systems—Specification for Design, Development, Production, Installation, and Servicing.*
2. ANSI/ISO/ASQC Q9002:1994 (ISO 9002) *Quality Systems—Specifications for Production, Installation and Servicing.*

3. ANSI/ISO/ASQC Q9003:1994 (ISO 9003) *Quality Systems—Specifications for Final Inspection and Test.*

In practice, these parts of the series are "nested," i.e. ISO 9001 contains all that is in 9002 and 9003, with some extra content. It is important to realize that these standards are not all different, but variations on one theme. The overall series is commonly referred to as "ISO 9000"; it is available from ANSI/ASQ and other sources. These standards have two key features. First, they are all about *documentation*. This means that everything written in the standard refers to specific documents—the scope of documentation is very wide. This does not mean that they don't have an effect on the product or service produced by a company; merely that these are not controlled directly by what is mentioned in the standard. Second, ISO 9000 is about the effectiveness of a quality *management* system—it does not impinge directly on the design, or the usefulness, or the fitness for purpose of the product. It is a quality *management* standard, not a product conformity standard. It is, therefore, entirely possible for a manufacturer with a fully compliant ISO 9000 system installed and working, to make a product that is not suited to its market (see Fig. 14.1).

Fig. 14–1

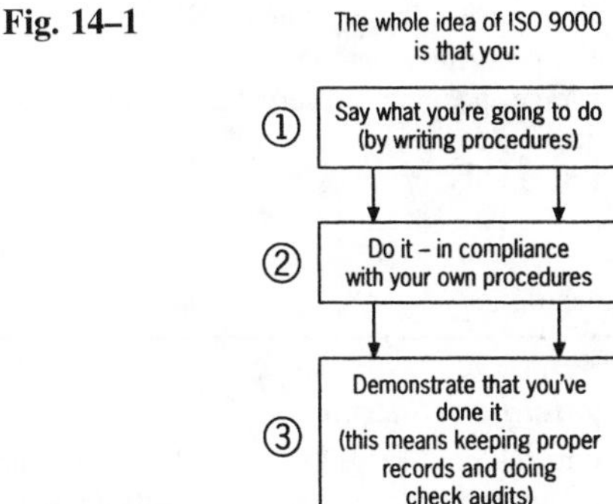
The whole idea of ISO 9000 is that you:
①
Say what you're going to do (by writing procedures)
②
Do it – in compliance with your own procedures
③
Demonstrate that you've done it (this means keeping proper records and doing check audits)

The Contents

The contents of ISO 9001 are divided into five discrete sections—0: "Introduction"; 1: "Scope"; 2: "References"; and 3: "Definitions"—that are all short and serve mainly as the introduction. The main part is Section 4, which is divided into 20 clauses, some of which are in turn split into several subclauses. These 20 clauses (see Table 14.1) form the core content of a good quality management system—i.e., *quality system elements.*

Table 14.1 ISO 9000 Quality System Elements

4.1	Management responsibility
4.2	Quality system
4.3	Contract review
4.4	Design control
4.5	Document and data control
4.6	Purchasing
4.7	Control of customer-supplied product
4.8	Product identification and traceability
4.9	Process control
4.10	Inspection and testing
4.11	Control of inspection, measuring, and test equipment
4.12	Inspection and test status
4.13	Control of non-conforming product
4.14	Corrective and preventive actions
4.15	Handling, storage, packing, preservation, and delivery
4.16	Control of quality audits
4.17	Internal quality audits
4.18	Training
4.19	Servicing
4.20	Statistical techniques

14.1.3 Quality System Certification

Most businesses that install an ISO 9000 system do so with the objective of having it checked and validated by an outside body. This is called *certification* (see Fig. 14.2). Certification bodies in some countries are themselves *accredited* by a national body, which ensures that their management and organizational capabilities are suitable for the task. The US Registrar Accredita-

Fig. 14–2

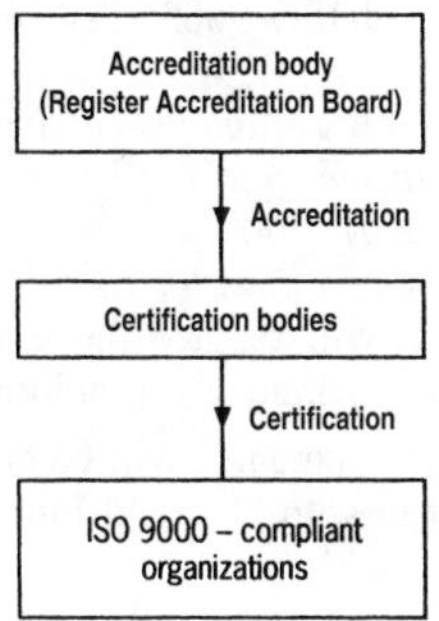

tion Board (RAB) is an ASQ affiliate that assesses suppliers and operates a program for certification of quality systems auditors.

QS-9000 has its own system of qualified registrars and an inventory of more than 12000 certified companies.

14.1.4 Future Developments: The "New Format" ISO 9000: 2000

ASQ is involved in the development of quality standards via the U.S. Technical Advisory Group for ISO Technical Committee 176. The existing structure of ISO 9000 standards is due to be replaced in 2000/2001.

Objectives of the Changes

- To make the requirements of the standards *easier to implement* in all businesses, not just manufacturing companies.
- To place more emphasis on *continual improvement* rather than simply to comply with the minimum requirements of an unchanging standard.
- To refocus on the idea of an overall quality *management* system rather than one that is limited to only quality assurance activities.

What Will Be in the New Standards?

The final structure and content of the standards is still being developed but the general pattern is likely to be as follows:

- ISO 9001: 2000 *Quality Management Systems* will replace ISO 9001: 1994, ISO 9002: 1994 and ISO 9003: 1994.

- ISO 9002: 1994 and ISO 9003:1994 will be withdrawn as separate standards.
- ISO 9004: 2000 *Quality Management Systems—Guidance for Performance Improvement* will give in-depth detail and explanation of ISO 9001: 2000.
- ISO 19011: 2000 *Guidelines for Auditing Quality and Environmental Systems* will replace the existing standard ISO 10011 covering the techniques of auditing.

It is feasible that these standards will be given additional numbers and titles to fit in with U.S. and European harmonization activities.

What Are the Implications?

- **A "single assessment standard."** All companies will be assessed against the "general" quality management standard ISO 9001: 2000. There will no longer be separate standards for companies that are involved only in design, testing, servicing, etc. This means that assessments may need to be *selective* i.e., only assessing the sections of a company's standard that are relevant to its activities.
- **Some registration changes.** Companies already certified to ISO 9001: 1994 will probably keep their existing registration number, but those previously certified to ISO 9002: 1994 and ISO 9003: 1994 will need to have their scope redefined.
- **All companies will need to refocus**—because of significant change in emphasis of the content of ISO 9001: 2000.

Table 14.2 summarizes the likely format.

Table 14.2 The "New Format" ISO 9001: 2000

The new format is likely to contain sections on:

- **The scope,** and its compatibility with the management systems.
- **Terms and definitions.**
- **Quality Management System (QMS) requirements**—general statements about what a QMS is, and is supposed to do.
- **Management responsibility**—a much expanded version of clauses 4.1 and 4.2 of ISO 9000: 1994, incorporating aspects such as legal and customer requirements, policy and planning objectives.

Table 14.2 The "New Format" ISO 9001: 2000 (cont.)

- **Resource management**—covering staff training (ex clause 4.18) and new requirements relating to competence, company infrastructure, and the work environment.
- **Product/Service "realization":** This section reflects the bulk of the content of clauses 4.3 - 4.17 of ISO 9000: 1994, but with a lot of changes and restructuring.
- **Measurement, analysis, and improvement:** This incorporates the philosophy of controlling non-conformities and the quality system itself by using reviews and audits (ISO 9001: 1994 clauses 4.13, 4.14, 4.16, 4.17), but extends the scope to include other, more detailed aspects of measurement and monitoring. It includes greater emphasis on continuous improvement.

14.2 Total Quality Management (TQM)

14.2.1 What is TQM?

TQM is a generic term for a blend of quality philosophies and management techniques in the areas of employee motivation, measurement, and reward. The four core concepts are:

Continuous Process Improvement

- Work is considered to be part of a continuous process. Improvement is seen as being made in a series of incremental steps.
- The improvement process is implemented from the lower working levels of an organization
- Processes are described using flowcharts.

Customer Focus

- Every work group focuses on providing value to the people who use its product.
- The needs of the customer are seen as an important aspect of quality control.

Defect Prevention

Prevention of defects is seen as being more important than sorting them out after they occur.

Universal Responsibility

This is a philosophy of seeing quality as being the responsibility of everyone, not just inspection or QA/QC departments.

14.2.2 Taguchi Methods

Doctor Genichi Taguchi developed a particular blend of quality concepts that started in Japan's telecommunications industry. Taguchi is a specific type of process control. It moves away from the estimation or counting of defective components to a wider view that encompasses *reducing* the variability of production, and hence the cost of defective items. The key points of the Taguchi idea are:

- Choose a manufacturing system or process that *reduces variability* in the end product.
- Use the standpoint of costs to choose design tolerances—asking what is an acceptable price to pay for a certain set of tolerances.
- Push the quality assessment back to the *design stage*—again, the objective is to reduce the possible variability of the product.

Taguchi's basic principles are not, in themselves, new. Many of the principles coincide with the requirements for good, practical engineering design.

14.3 Statistical Process Control (SPC)

SPC is a particular type of quality control used for mass production components such as nuts and bolts, motor and auto components, etc. It relies on the principle that the pattern of variation in dimension, surface finish, and other manufacturing "parameters" can be studied and controlled by using *statistics*.

14.3.1 Normal Distribution

The key idea is that by inspecting a sample of components, it is possible to infer the compliance (or non-compliance) with specification of the whole batch. The core assumption is that of the *normal distribution* (see Fig. 14.3).

Fig. 14–3

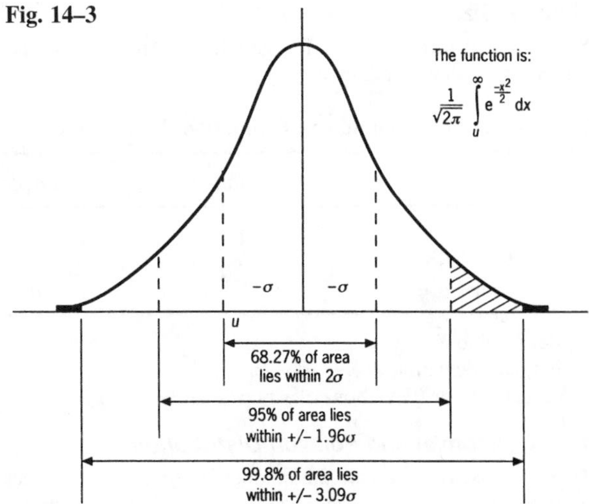

The quantities used are:

Standard deviation, $\sigma = \sqrt{\text{variance}}$

$$\sigma = \sqrt{\frac{f_1(x_1 - \overline{x})^2 + f_2(x - \overline{x})^2 + \ldots}{N}}$$

N = number of items
f = frequency of items in each group
x_1, x_2, etc. = mid size of the groups
$\overline{x}$ = arithmetic mean

From the normal distribution, a "rule of thumb" is :

1 in 1000 items lie outside $\pm 3\sigma$
1 in 40 items lie outside $\pm 2\sigma$

Sample Size

Symbols and formulae used for sample and "population" parameters are shown in Table 14.3.

Table 14.3 Sample and Population Parameters

	Population	*Sample*
Average value	$\overline{X}$	$\bar{x}$
Standard deviation	σ	s
No. of items	N	n

Mean value $\overline{X} = \bar{x}$
Standard deviation of $\bar{x} = \sigma/\sqrt{2\pi}$
Standard error (deviation) of $s = \sigma/\sqrt{2n}$

14.3.2 Binomial and Poisson Distributions

These are used to estimate the number (p) of defective pieces or dimensions. The easiest method is to use a Poisson distribution that is based on the exponential functions e^x and e^{-x}

$$e^{-x}e^{x} = e^{-x} + xe^{-x} + \frac{x^2e^{-x}}{2!} + \frac{x^3e^{-x}}{3!} + \ldots$$

This provides a close approximation to a binomial series and gives a probability of there being *less than* a certain number of defective components in a batch.

14.3.3 Control Charts

Control charts are a statistical tool for analyzing the state of an industrial process, i.e., whether or not it is under control. There are two main types: *Range* charts and *Average* charts.

Range Charts

The range chart shows the readings (of size, etc.) compared to a centerline for the range R of readings and their upper and lower control limits (UCL, LCL). Figure 14.4 shows the difference between processes that are under control and not under control.

Fig. 14–4

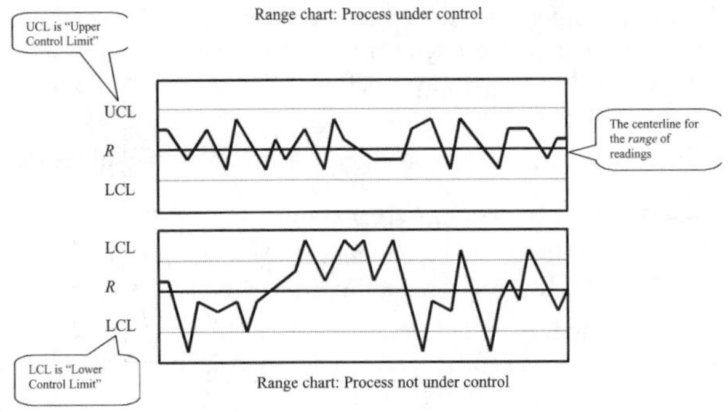

Average Charts

The average chart (see Fig. 14.5) shows readings compared to a centerline for the average X_m of the readings.

Fig. 14–5

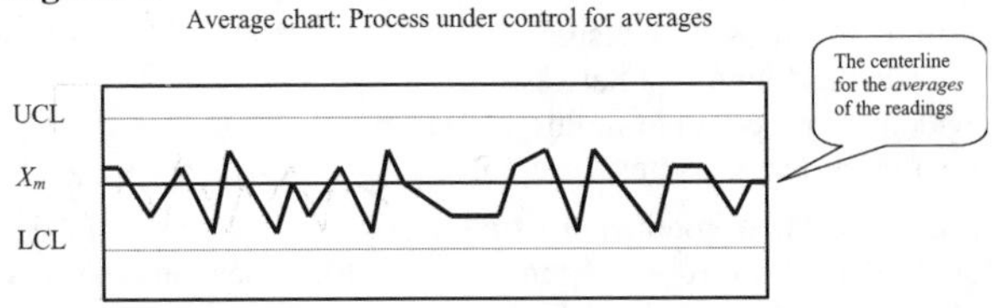

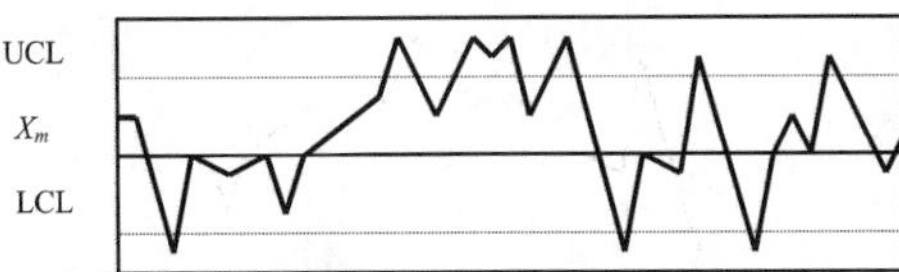

14.4 Reliability

It is not straightforward to measure, or even define, the reliability of a completed engineering component. It is even more difficult at the design stage, before a component or assembly has been manufactured.

- In essence reliability is about *how*, *why*, and *when* things fail.

14.4.1 The Theoretical Approach

There is a well-developed theoretical approach based on probabilities. Various design-reliability tools are used, such as:

- Failure Modes and Effects Analysis (FMEA)
- Failure Modes, Effects, and Criticality Analysis (FMECA)
- Mean Time To Failure (MTTF)
- Mean Time between Failures (MTBF)
- Fault Tree Analysis (FTA)
- Monte Carlo Analysis (based on random events)

Failure Modes and Effects Analysis (FMEA)

FMEA is a qualitative tool used to evaluate the reliability of engineering products and systems. It can also be used to:

- find weaknesses in design.
- evaluate design alternatives.
- identify all possible failure modes and their effects.
- suggest necessary test programs.

FMEA is often conducted using worksheets, which hold detailed data about sub-components, their functions, and the way in which each is likely to fail (see Fig. 14.6).

Fig. 14–6

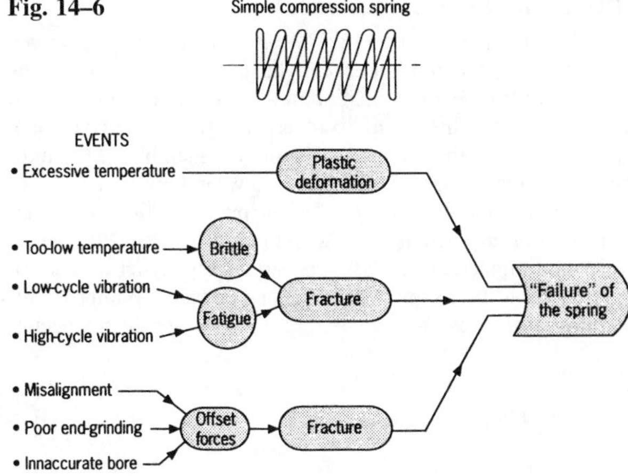

The principle of failure mode effects analysis (FMEA)

MTTF and MTBF

Mean time to failure (MTTF) is defined as the mean operating time between successive failures without considering repair time. Mean time between failures (MTBF) includes the time needed to repair the failure. If a component or system is not repaired, then MTTF and MTBF are equal (see Fig. 14.7).

Fig. 14–7

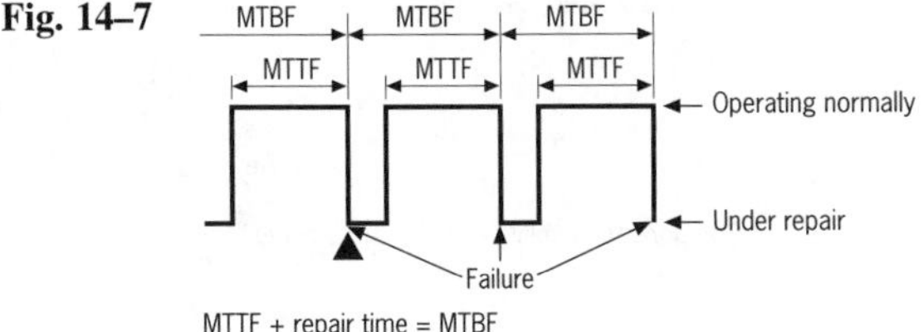

14.4.2 The Practical Approach

The "bathtub curve" (see Fig. 14.8) is surprisingly well proven at predicting when failures can be expected to occur. The chances of failure are quite high in the early, operational life of a product item; this is due to inherent defects or fundamental design errors in the product, or incorrect assembly of the multiple component parts. A progressive wear regime then takes over for the "middle 75 percent" of the product's life—the probability of failure here is low. As lifetime progresses, the rate of deterioration increases, causing progressively higher chances of failure. Table 14.4 shows steps that can be taken to improve reliability at the design stage, before reliability problems occur.

Fig. 14–8

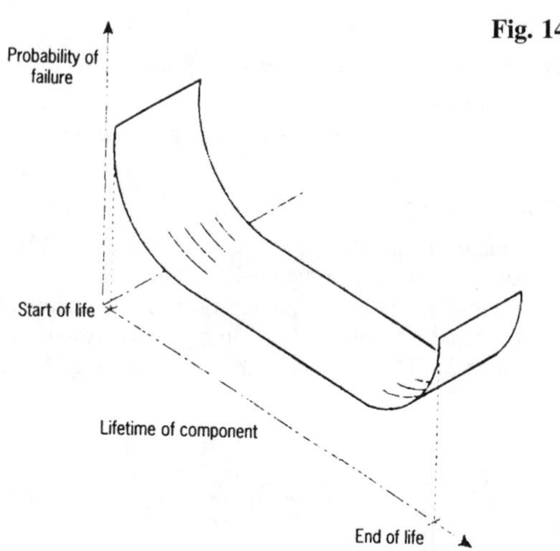

Component reliability – the "bathtub curve"

Table 14.4 Improving Design Reliability: Main Principles

- *Reduce static loadings*: It is often the most highly stressed components that fail first.
- *Reduce dynamic loadings*: Dynamic stress and shock loadings can be high.
- *Reduce cyclic conditions*: Fatigue is the largest single cause of failure of engineering components.
- *Reduce operating temperature*: Operation at near ambient temperatures improves reliability.
- *Remove stress raisers*: They cause stress concentrations.
- *Reduce friction*: Or keep it under control.
- *Isolate corrosive and erosive effects*: Keep them away from susceptible materials.

14.4.3 Design For Reliability—A New Approach

Design for reliability (DFR) is an evolving method of stating and evaluating design issues in a way that helps achieve maximum reliability in a design. The features of this "new approach" are:

- It is a quantitative but *visual* method—so not too difficult to understand.
- No separate distinction is made between the functional performance of a design and its reliability—both are considered equally important.
- It does not rely on pre-existing failure rate data (which can be inaccurate).

The Technique

Design parameters are chosen with the objective of maximizing all of the safety margins that will be built-in to a product or system. All the possible modes of failure are investigated and then expressed as a set of design constraints (see Fig. 14.9). The idea is that a design which has the highest safety margin with respect to all the constraints will be the most reliable (point X in the figure). Constraints are inevitably defined in a variety of units so a grading technique is required, which yields a non-dimensional performance measure of each individual constraint.

Fig. 14–9

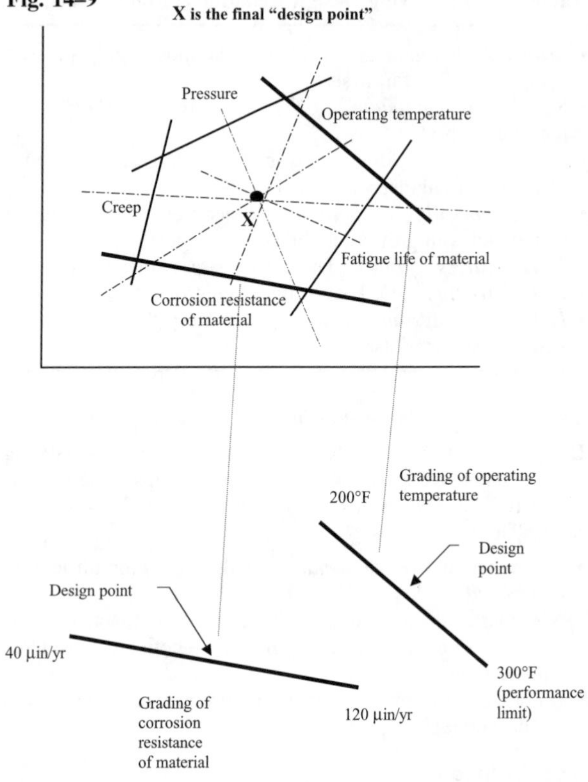

14.5 Modern "System" Tools

Over the past ten years, the "mission" of many large engineering corporations has been centered around the objective of improving the efficiency of data and resource management. Rapidly developing technologies such as Electronic Data Exchange, PC-based software packages, and the Internet have turned this into one of the fastest moving areas of manufacturing and project management. The two main generic areas are Product Data Management and Enterprise Resource Planning.

14.5.1 Product Data Management (PDM)

The Objective of PDM: To create a single place for all the information about a company's products and services, and to keep this information up-to-date as things change (see Fig. 14.10). The main computerized techniques of PDM are, therefore:

- Accurate dissemination of engineering data involved inside and outside the organization.
- Timely compilation of engineering data as it develops.
- "Automation" of the design management process.

Fig. 14–10

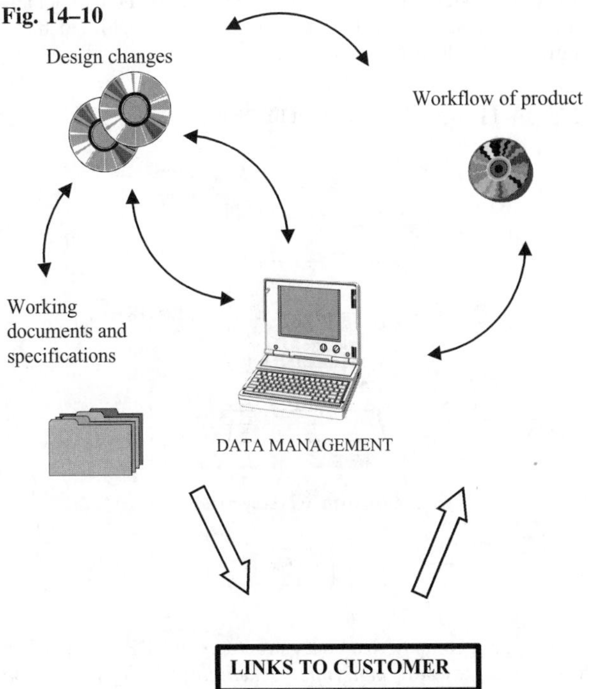

The activities of PDM are therefore orientated toward the *function* of a product (or project) and how it is configured. The idea is to produce data that can be effectively managed and used downstream of the engineering function so it involves "data-*push*."

14.5.2 Enterprise Resource Planning (ERP)

The Objective of ERP: To enable the planning of resources at the strategic, tactical, and operational levels of an engineering project or enterprise. This means that ERP is about implementing consistent data structures in all parts of an organization (see Fig. 14.11). This can be achieved using several software models, all of which have the common goal of achieving *consistency* of the type and form of data held and transferred in an electronic form.

Fig. 14–11 THE PRINCIPLES OF ERP

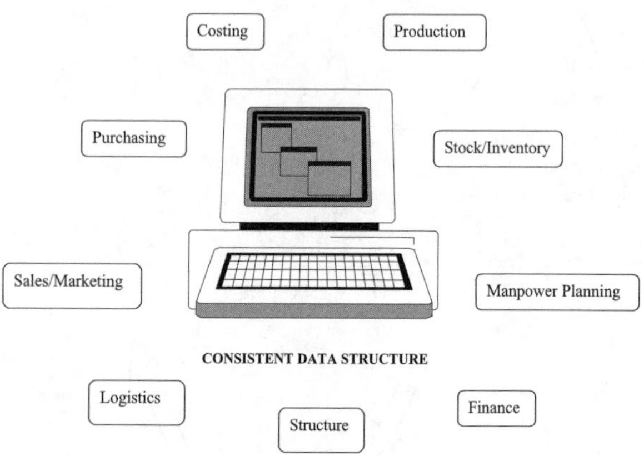

The main computerized data structures in ERP are therefore orientated toward *materials scheduling and production* processes, i.e., data structure that define parts of a product (or project) and how they are assembled in the works (or on-site).

- Remember the difference:

  PDM deals with *function and configuration.*
  ERP deals with *parts and their assembly.*

14.6 Useful References

QS-9000 Web sites

American Society of Quality (ASQ): www.asq.org/standcert/9000.html for an introduction to QS-9000 and a list of useful links to:

- QS-9000 technical details
- Certified company directory
- Registrar Accreditation Body listings

ISO 9000 Web sites

American Society of Quality (ASQ): www.qualitypress.asq.org/perl/catalog to browse publications relating to ISO 9000
Registrar Accreditation Board (RAB): www.rabnet.com
US standards group: www.standardsgroup.asq.org
International Organization for Standardization (ISO online): www.iso.ch/welcome.html
European Foundation for Quality Management: www.efqm.org
Quality Network (UK): www.quality.co.uk

Useful TQM References

1. Hradesky. J. *Total Quality Management Handbook*:1995. Pub: ASQ. ISBN 0-07-030511-0.
2. Taguchi, G. *Experimental Designs*, 3rd edition, 1976. Pub: Marmza Publishing Company, Tokyo.

Useful SPC Standards

1. ANSI/ASQC Z1.4: 1993 *Sampling Procedures and Table for Inspection by Attributes.*
2. ANSI/ASQC S2: 1995 *Introduction to Attributes Sampling.*
3. ANSI/ISO/ASQ A 3534-1: 1993 *Probability and General Statistical Terms.*
4. ANSI/ASQC B1: 1996 *Guide for Quality Control Charts.*
5. ANSI/ASQC B2: 1996 *Control Chart Method of Analyzing Data.*
6. ANSI/ASQC B3 *Control Chart Method of Controlling Quality during Production.*

SPC Websites

Using Control Charts: www.outboundtrain.com/html/spc_control.html

For a list of statistical-related production standards: http://standardsgroup.asq.org/about/s/statlist.htm

FMEA Reference Standards

The U.S. Department of Defense and ANSI publish standards covering FMEA: e.g., MIL-STD-A8 1629: *Procedures for Performing a FMEA/FMECA Analysis* See also*: Mechanical Engineer's Handbook* (Chapter 20). 2nd edition 1998. Myer Kutz. Pub: John Wiley. ISBN 0-471-13007-9

Reliability Standards

The U.S. Department of Defense produces "MIL standard" documents, which are also utilized in many non-defense design specifications. Key ones relating to reliability engineering are:

1. MIL-STD-756 *Reliability Modeling and Prediction.*
2. MIL-STD-721 *Definitions of Terms for Reliability and Maintainability.*
3. MIL-STD-781 *Reliability Design Qualifications and Production Acceptance Tests.*
4. ANSI/IEC/ASQ D3000-1-1:1997 *Analysis Techniques for Dependability: Guide on Methodology.*
5. MIL-HDBK-472 *Maintainability Prediction.*
6. ISO 2382-14: 1978 *Reliability, Maintainability, and Availability.*

Useful Reliability References

1. Bentley. J.P *An Introduction to Reliability Engineering*: Pub: John Wiley Inc. 1995. ISBN 0-471-01833-3
2. Krishnamoorthi. K. S. *Reliability Methods for Engineers*: 1992. ISBN 0-87389-161-9
3. J. W. Wilber and N. B. Fuqua: *A Primer for DOD Reliability.* Maintainability and safety standards document No PRIM 1: 1988. Rome Air Development Center, Griffiths Air Force Base, Rome, NY.
4. Shooman, M. L., *Probabilistic Reliability—An Engineering Approach*: 1990 Pub: R. E. Krieger, Melbourne.

Reliability Web Sites

National Information Center for Reliability Engineering: www.enre.umd.edu/

NASA Engineering Analysis Branch: http://ixeab3.larc.nasa.gov/ateb.html

PDM Web Sites

CIO Executive Research Center: www.cio.com/forms/data/

ERP Web Sites

1. The ERP/Supply Chain Research Center: www.cio.com/forums/erp.
2. 26-page paper "The future of ERP" from Quantiv Inc: www.quantiv.com

Section 15

Project Engineering

15.1 Project Planning

The most common tool to help plan and manage a project is the Program Evaluation and Review Technique (PERT). In its simplest form it is also known as Critical Path Analysis (CPA) or network analysis. It is used for projects and programs of all sizes and marketed as software packages under various trade names. The technique consists of five sequential steps (see Fig. 15.1).

Fig. 15–1

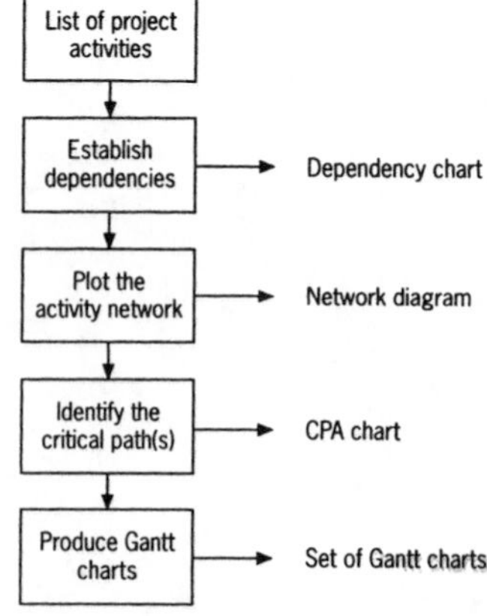

15.1.1 Listing Activities

All individual activities are input into the package. There may be thousands of these on a large construction program.

15.1.2 Tabulating Dependencies

The dependency table is the main step in organizing the logic of the listed activities. It shows the previous activities on which each individual activity is dependent (see Fig. 15.2).

Fig. 15–2

No.	Activity: e.g.	Preceding activity
1	Conceptual design	–
2	Embodiment design	1
3	Detailed design	2
4	Research materials	–

Dependency table

15.1.3 Creating a Network

A network is created showing a graphical "picture" of the dependency table (see Fig. 15.3). The size of the boxes and length of interconnecting lines have no program significance. The lines are purely there to link dependencies, rather than to portray timescale.

Fig. 15–3

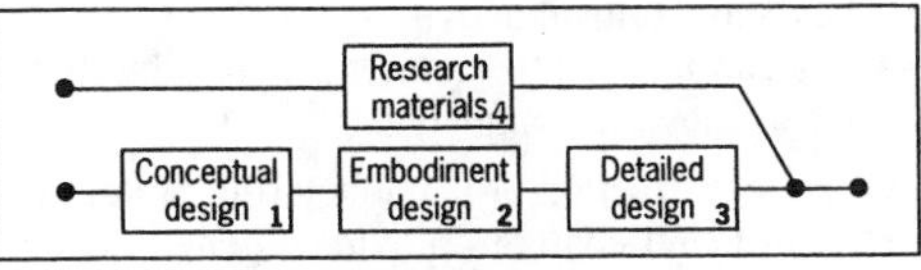

Network diagram

15.2 Critical Path Analysis (CPA)

The CPA introduces the concept of timescale into the network. It shows not only the order in which each project activity is undertaken, but also the duration of each activity. CPA diagrams are traditionally shown as a network of linked circles, each containing the three pieces of information shown. The critical path is shown as a thick arrowed line and is the path through the network that has *zero float*. Float is defined as the amount of time an activity can shift, without affecting the pattern or completion date of the project (see Fig. 15.4).

Fig. 15–4

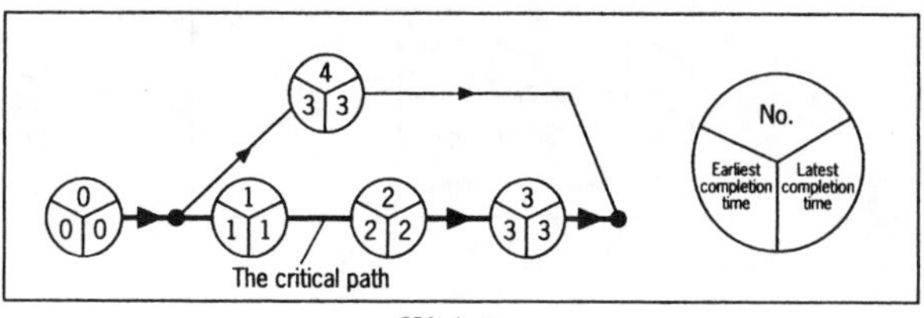

CPA chart

15.3 Planning With Gantt Charts

Gantt charts are produced from the CPA package and are used as the standard project management documents (see Fig. 15.5). Their advantages are :

—they provide an easy-to-interpret picture of the project;
—they show critical activities;
—they can be used to monitor progress by marking off activities as they are completed.

A large construction or manufacturing project will have a hierarchy of Gantt charts to provide a general overview and more detailed analysis of the important parts of the project.

Fig. 15–5

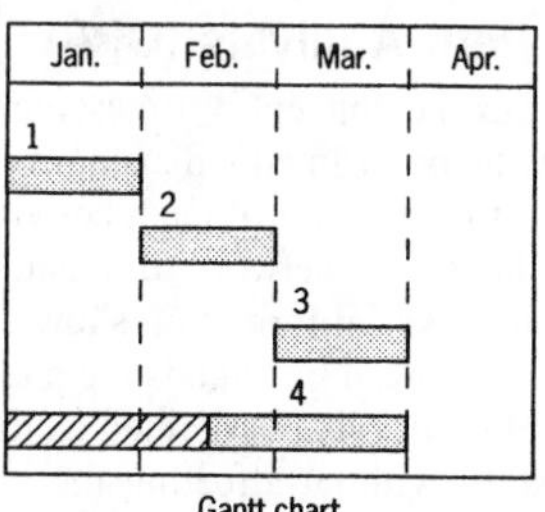

Gantt chart

15.4 Linear Programming

Linear programming is a mathematical tool normally computerized, employed to determine the best use of a set of available resources. It is a systematic procedure that can help with decision-making. It works using the principle that all processes are governed by a set of multiple constraints that restrict what can be done. Figure 15.6 shows a graphical example for the organization of an engineering construction project.

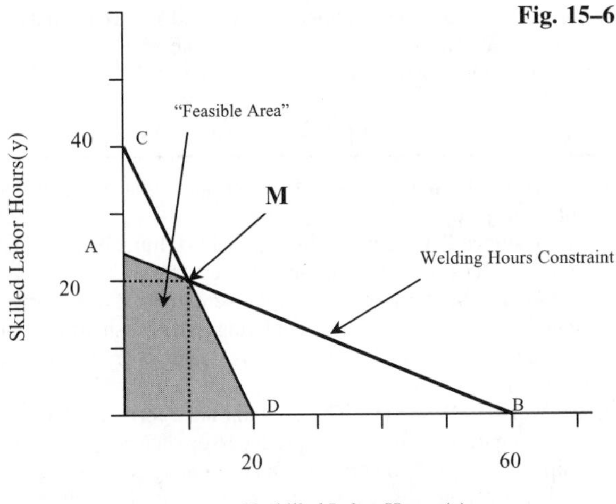

15.4.1 Methodology (see Fig. 15.6)

1. The objective is to maximize the value of work done, which is given by $\$ = 3x + 4y$.
2. Line AB indicates a constraint on welding time, which follows the form $2x + 5y \leq 120$ hrs per week.
3. Line CD indicates a constraint on site access time, which follows the form $4x + 2y \leq 80$ hrs per week.
4. The shaded area represents the "feasible area" within which it is possible to satisfy both sets of imposed constraints.
5. The maximum value of work done is achieved at point **M**, at the corner of the "feasible area."

This graphical method of linear programming can be used only with two sets of constraints. A tabular method (called the *simplex* technique) normally forms the basis of computerized linear program packages and can handle more complex situations.

15.5 Project Management Software Packages

PC-based project management and scheduling packages are produced by numerous software manufacturers. All have similar objectives: to handle numerous data and report formats in the most effective way, and follow the basic structure of project information "views," as shown in Table 15.1

Table 15.1 Project Information "Views"

1. **A Project listing:** A list of sub-projects that make up the main project.
2. **A Resource list view:** A spreadsheet of human and material resources, their availability, and costs.
3. **A Gantt chart view:** The representation of the project schedule as series of rows depicting start/finish time, task interdependency, and current status. Gantt chars are shown to an (accurate) timescale.
4. **Network/CPA view:** The critical path "derivation" of the Gantt chart showing fixed predecessor/successor relationships that exist in the project. On-line changes can be made, which are then reflected in the corresponding Gantt chart view. Most packages can display CPA details of sub-projects, showing how they integrate into the main project CPA display.
5. **Time and cost graphs:** Displayed as cumulative histograms or line charts, these show cost and time performance against budget.
6. **Report output:** Read-only reports of any of the above functions. These are used as essential "feedback" to project management and staff.

Virtually all modern packages adapt easily for use across multi-site computer networks and so can handle projects spread over several different locations. Many now include web browsers for internet access. Smaller packages of "Project Communications" software are used for specific communication purposes between project managers and their teams. They deal with specific items, such as scheduled tasks, work assignments, and costs.

15.6 Rapid Prototyping

The later stages of the design process for many engineering products involve making a prototype. A prototype is a non-working (or sometimes working) full-size version of the product under design. Despite the accuracy and speed of CAD/CAM packages, there are still advantages to be gained by having a model in physical form, rather than on a computer screen. Costs, shapes, colors, etc. can be more easily assessed from a physical model.

The technology of *rapid prototyping* (RP) produces prototypes in a fraction of the time, and cost, of traditional techniques using wood, card, or clay models. Quickly available, solid prototypes enable design ideas to be tested and analyzed fast, hence increasing the speed and efficiency of the design process.

15.6.1 Prototyping Techniques

These are state-of-the-art technologies, which are developing quickly. Most use similar principles of building up a solid model by stacking together elements or sheets. The main ones are:

—*Stereolithography*: This is the most common method. It involves laser solidification of a thin polymer film, which is floating on a bath of fluid. Each layer is solidified sequentially, the shapes being defined by the output from a CAM package.

—*Laser sintering*: Here a CAM-package driven laser is used to sinter the required shape out of a thin sheet of powder.

—*Laminated manufacture*: This is a simpler version of the same principle. Laminated sheets of foam are stuck together in an automated process using adhesive or heat.

15.7 Value Analysis

Value analysis (or value engineering) is a generic name relating to quantifying and reducing the cost of an engineering product or project. Value analysis is about asking questions at the design stage, before committing to the costs of manufacture. All aspects of product design, manufacture, and operation are open to value analysis. Several areas tend to predominate: shape, materials of construction, surface finish, and tolerances (see Fig. 15.7).

Fig. 15–7

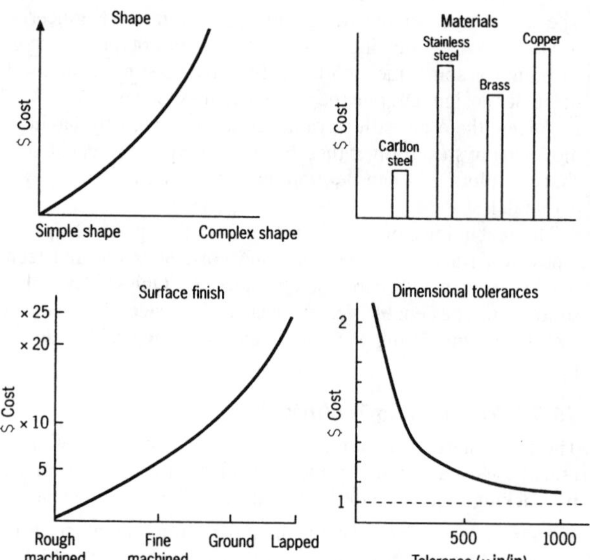

15.8 Useful References

Web Sites: Project Management

Generic information and information/links to software package manufacturer is available at The US Project Management Institute: www.pmi.org/

Standards: Project Management

1. ASSIST Standard NFGS-01321A: 1996. *Network Analysis Schedules.*
2. AIA Standard 1.12: 1998. *Project Management.*
3. IEEE Standard 1058: 1998. *Standard for Software Project Management Plans.*
4. ANSI Z94.4: 1998. *Cost Engineering and Project Management.*
5. ISO 10006:1997. *Guidelines to Quality in Project Management*

Standards: Rapid Prototyping

1. ASTM E1340: 1994. *Standard Guide for Rapid Prototyping of Computerized Systems.*

Web Sites: Rapid Prototyping

For a table of RP links and several dedicated RP web crawlers: www.pitt.edu/~roztocki/rapidpro/rapidpro.htm

A list of RP sites is also available at www.biba.uni-bremen.de/groups/rp/rp_sites.html

Standards: Value Analysis

1. ASTM E1804: 1994. *Standard Practice for Performing and Reporting Cost Analysis.*
2. ACPA Standard PC2298: 1991. *Life Cycle Cost Analysis.*

Web Sites: Value Analysis

For a listing of value analysis links: www.value-analysis.com/weblinks.htm

See also:

Greasly. A. *Project Management for Product and Service Improvement*: 1997: Pub Butterworth. ISBN 0-7506-3764 2.

Section 16

Computer Aided Engineering

Computer Aided Engineering (CAE) is the generic name given to a collection of computer-aided techniques used in mechanical engineering:

Computer Aided Engineering (CAE) comprises:

- **CAD: Computer Aided Design (or Drafting)**
  —*Computer Aided Design* is the application of computers to the conceptual/design part of the engineering process. It includes analysis and simulation.
  —*Computer Aided Drafting* is the application of computer technology to the production of engineering drawings and images.
- **CAM: Computer Aided Manufacture** relates to the manufacture of a product using computer-controlled machine tools of some sort.
- **MRP: Materials Requirements Planning/Manufacturing Resource Planning** defines when a product is made and how this fits in with the other manufacturing schedules in the factory.
- **CIM: Computer Integrated Manufacture** is the integration of all the computer-based techniques used in the design and manufacture of engineering products.

Figure 16.1 shows a general representation of how these techniques fit together.

Fig. 16–1

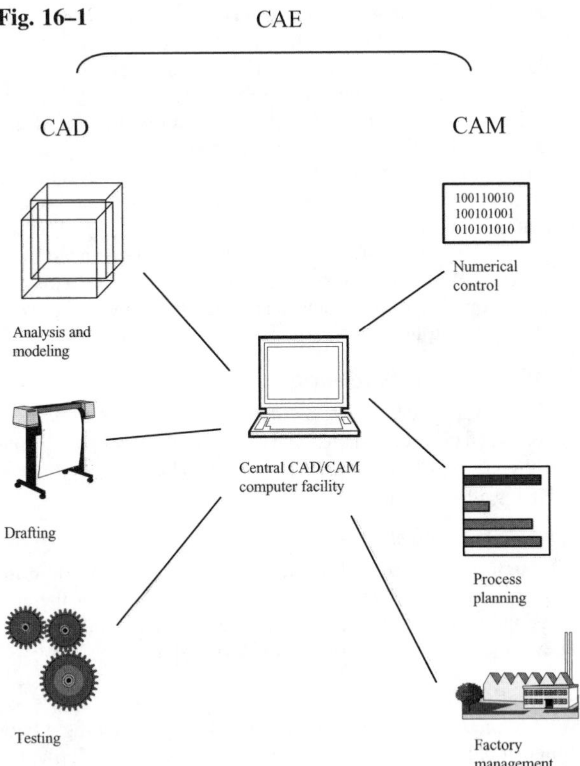

16.1 CAD Software

CAD software exists at several levels within an overall CAE system. It has different sources, architecture, and problems. A typical structure is:

- *Level A—Operating systems*: Some are manufacturer-specific and tailored for use on their own systems.
- *Level B—Graphics software*: This governs the type and complexity of the graphics that both the CAD and CAM elements of a CAE system can display.

- *Level C—Interface/exchange software*: This comprises the common software that will be used by all the CAD/CAM application, e.g., user interface, data exchange, etc.
- *Level D—Geometric modeling programs*: Most of these are designed to generate an output, which can be translated into geometric form to guide a machine tool.
- *Level E—Applications software*: This is the top level of vendor-supplied software and includes drafting and analysis/simulation facilities.
- *Level F—User-defined software*: Many systems need to be tailored before they can become truly user-specific. This category contains all the changes required to adapt vendor software for custom use.

16.2 Types of Modeling

CAD software packages are divided into those that portray two-dimensional or three-dimensional objects. 3-D packages all contain the concept of an *underlying model*. There are three basic types (as shown in Fig. 16.2).

Wireframe Models

Although visually correct, these do not contain a full description of the object. They contain no information about the surfaces and cannot differentiate between the inside and outside. They cannot be used to link to a CAM system.

Surface Models

Surface models are created (conceptually) by stretching a two-dimensional "skin" over the edges of a wireframe to define the surfaces. They can therefore define structure boundaries, but cannot distinguish a hollow object from a solid one. Surface models can be used for geometric assembly models etc., but not analyses, which require recognition of the solid properties of a body (finite element stress analysis, heat transfer, etc.).

Solid Models

Solid models provide a full three-dimensional geometrical definition of a solid body. They require large amounts of computer memory for definition and manipulation but can be used for finite element applications. Most solid modeling systems work by assembling a small number of "building block" reference shapes.

Fig. 16–2

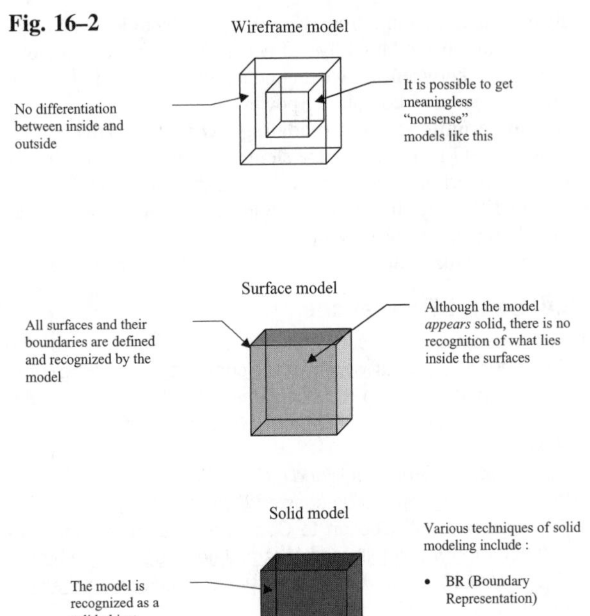
Wireframe model
No differentiation between inside and outside
It is possible to get meaningless "nonsense" models like this
Surface model
All surfaces and their boundaries are defined and recognized by the model
Although the model appears solid, there is no recognition of what lies inside the surfaces
Solid model
The model is recognized as a solid object
Various techniques of solid modeling include :
• BR (Boundary Representation)
• CSG (Constructive Solid Geometry)
• FM (Faceted Modeling)

16.3 Finite Element (FE) Analysis

FE software is the most widely used type of engineering analysis package. The basic idea is that large three-dimensional areas are subdivided into small triangular or quadrilateral (planar) or hexahedral (three dimensional) *elements,* then subject to solution of multiple simultaneous equations. The general process is loosely termed *mesh generation*. There are four types that fall into the basic category.

- **Boundary element modeling (BEM):** This is a simplified technique used for linear or static analyses where boundary conditions (often assumed to be at infinity) can be easily set. It is useful for analysis of cracked materials and structures.

- **Finite element modeling (FEM):** The technique involves a large number of broadly defined (often symmetrical) elements set between known boundary conditions. It requires large amounts of computing power.
- **Adaptive finite element modeling (AFEM):** This is a refinement of FEM in which the element "mesh" is more closely defined in critical areas. It produces better accuracy.
- **Finite difference method:** A traditional method, which has now been superseded by other techniques. It is still used in some specialized areas of simulation in fluid mechanics.

16.4 Useful References

1. For a general introduction to types of CAD/CAM, go to *The Engineering Zone* at www.flinthills.com/~ramsdale/EngZone/cadcam.htm. This site also contains lists of links to popular journal sites such as *CAD/CAM Magazine* and *CAE Magazine*.
2. *Finite Element Analysis World* includes listings of commercial software: www.comco.com/feaworld/feaworld.html.
3. For a general introduction to Computer Integrated Manufacture (CIM): www.flinthills.com/~ramsdale/EngZone/cim.htm.
4. *The International Journal of CIM*: www.tandfdc.com/jnls/cim.htm.
5. For an online introductory course on CIM: www.management.mcgill.ca/course/msom/MBA/mgmt-tec/students/cim/TEST.htm.
6. For a list of PDM links: www.flinthills.com/~ramsdale/EngZone/pdm.htm.
7. *The PDM Information Center* PDMIC is a good starting point for all PDM topics: www.pdmic.com/. For a bibliography listing: www.pdmic.com/bilbliographies/index.html.

Section 17

Welding

17.1 Welding Processes

17.1.1 Shielded Metal Arc (SMAW) or Manual Metal Arc (MMA)

This is the most commonly used technique. There is a wide choice of electrodes, metal, and fluxes, allowing application to different welding conditions. The gas shield is evolved from the flux, preventing oxidation of the molten metal pool (see Fig. 17.1).

Fig. 17–1

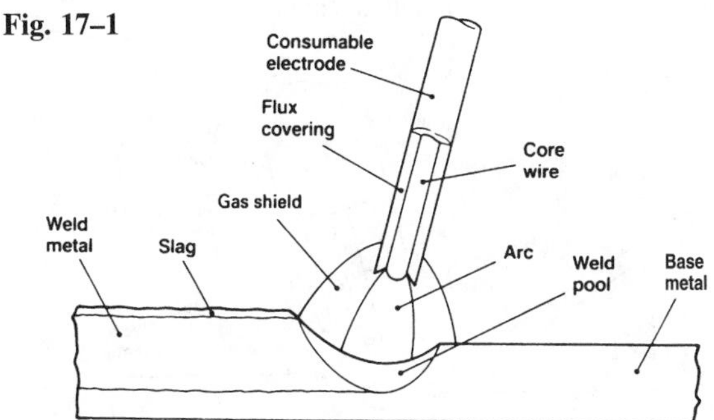

17.1.2 Metal Inert Gas (MIG)

Electrode metal is fused directly into the molten pool. The electrode is therefore consumed rapidly, being fed from a motorized reel down the center of the welding torch (see Fig. 17.2).

Fig. 17–2

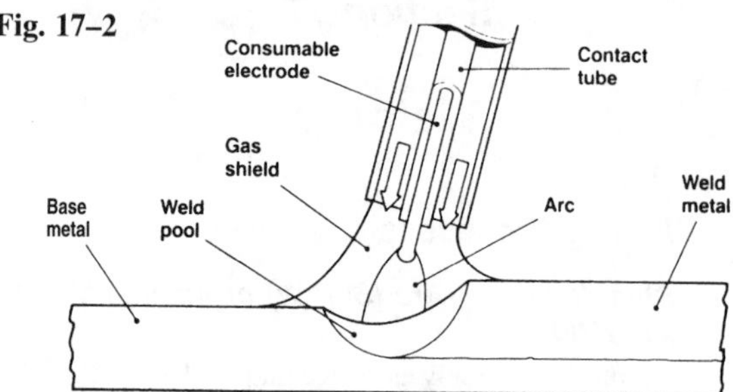

17.1.3 Tungsten Inert Gas (TIG)

This uses a similar inert gas shield to MIG but the Tungsten electrode is not consumed. Filler metal is provided from a separate rod fed automatically into the molten pool (see Fig. 17.3).

Fig. 17–3

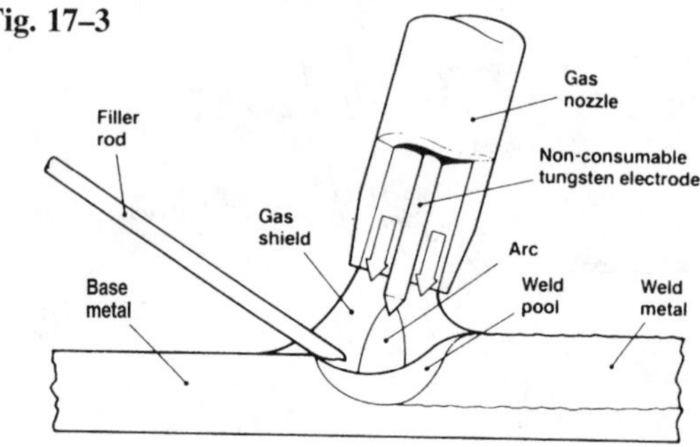

17.1.4 Submerged Arc Welding (SAW)

Instead of using shielding gas, the arc and weld zone are completely submerged under a blanket of granulated flux (see Fig. 17.4). A continuous wire electrode is fed into the weld. This is a common process for welding structural carbon or carbon–manganese steelwork. It is usually automatic, with the welding head being mounted on a traversing machine. Long continuous welds are possible with this technique.

Fig. 17–4

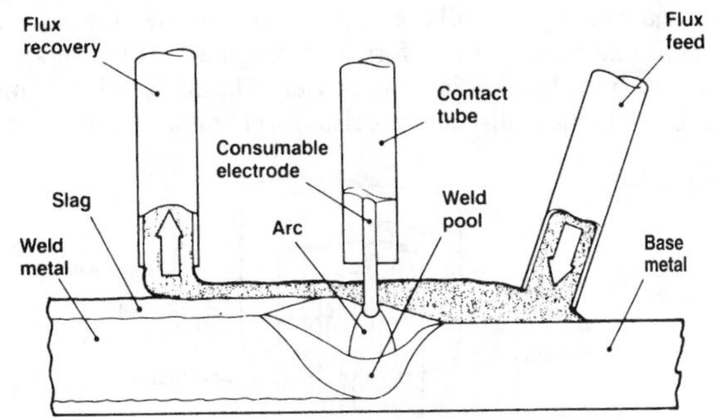

17.1.5 Flux-Cored Arc Welding (FCAW)

Similar to the MIG process, but this uses a continuous hollow electrode filled with flux, which produces the shielding gas (see Fig. 17.5). The advantage of the technique is that it can be used for outdoor welding, as the gas shield is less susceptible to draughts.

Fig. 17–5

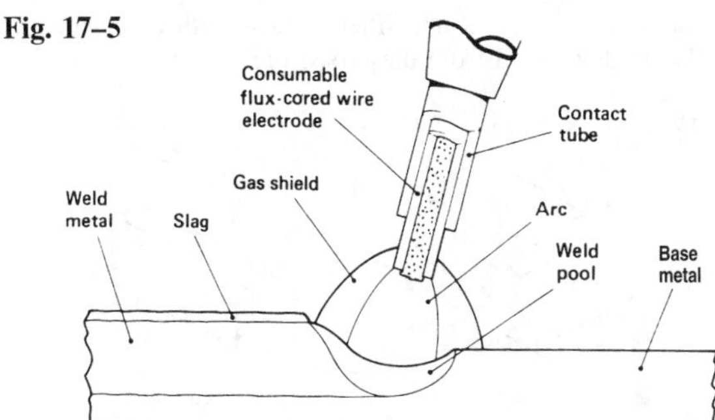

17.1.6 Electrogas Welding (EGW)

This is a mechanized electrical process using an electric arc generated between a solid electrode and the workpiece. It has similarities to the MIG process.

17.1.7 Plasma Welding (PW)

Plasma welding is similar to the TIG process (see Fig. 17.6). A needle-like plasma arc is formed through an orifice and fuses the base metal; shielding gas is used. Plasma welding is most suited to high-quality and precision welding applications.

Fig. 17–6

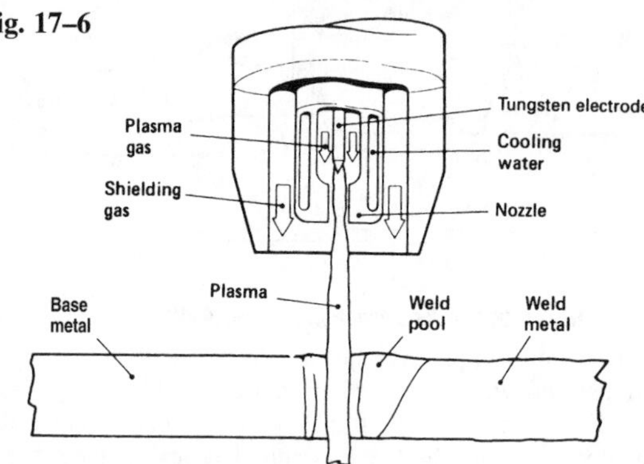

17.2 Weld Types and Orientation

The main *types* are butt and fillet welds—with other specific ones being developed from these (see Fig. 17.7).

Fig. 17–7

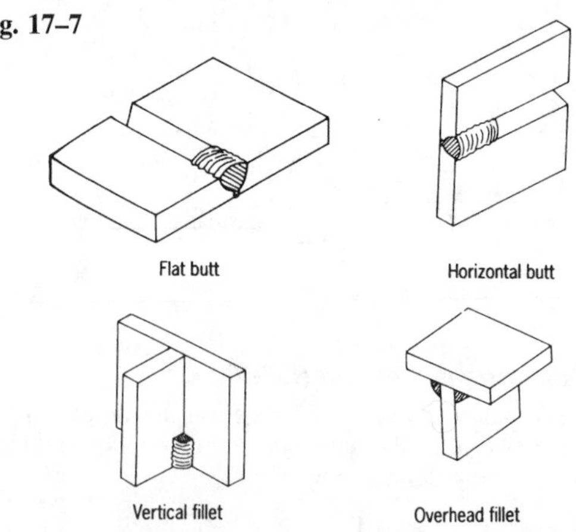

Orientation of the weld (i.e., the position in which it was welded) is also an important factor. Weld positions are classified formally in technical standards, such as AWS and *The ASME Pressure Vessel Code*, Sections II and IX.

17.2.1 Weld Terminology

Fillet and butt welds features have specific terminology that is used in technical standards such as AWS A2.4 (see Fig. 17.8).

Fig. 17–8

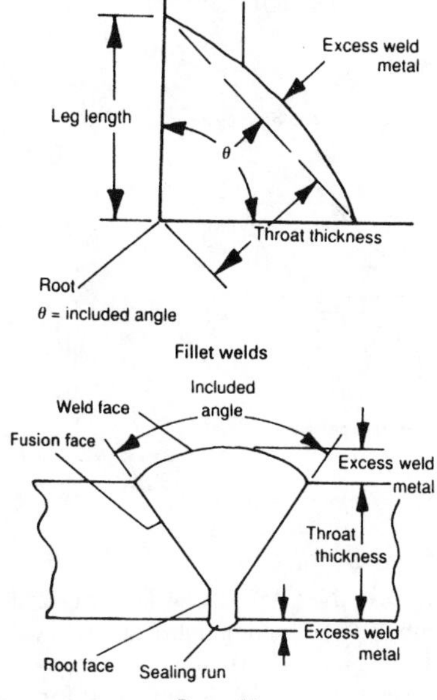

Fig. 17–8 (cont.)

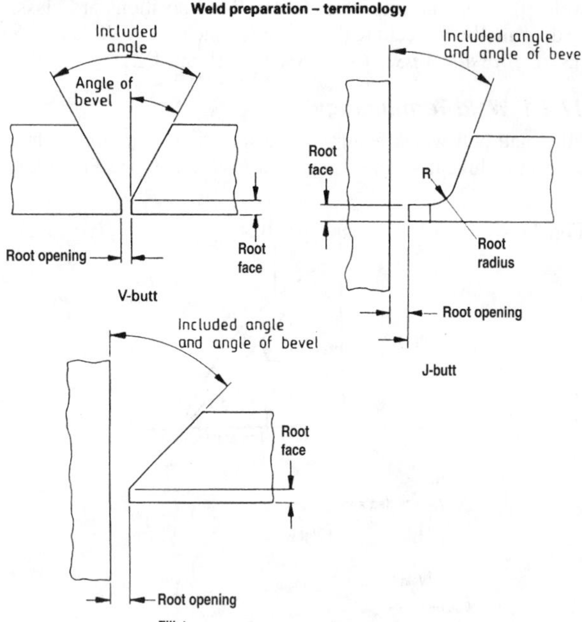

17.3 Welding Symbols

Standards such as ANSI/AWS A2.4 and ISO 2553 contain libraries of symbols to be used on fabrication drawings to denote features of weld preparations and the characteristics of the welds themselves (see Figs. 17.9 and 17.10 for examples).

Fig. 17–9

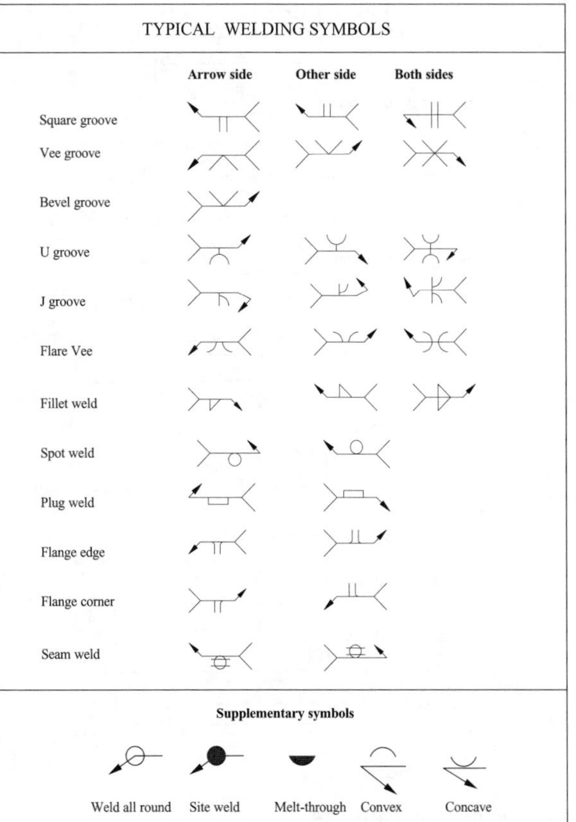
TYPICAL WELDING SYMBOLS
Arrow side
Other side
Both sides
Square groove
Vee groove
Bevel groove
U groove
J groove
Flare Vee
Fillet weld
Spot weld
Plug weld
Flange edge
Flange corner
Seam weld
Supplementary symbols
Weld all round
Site weld
Melt-through
Convex
Concave

Fig. 17–10

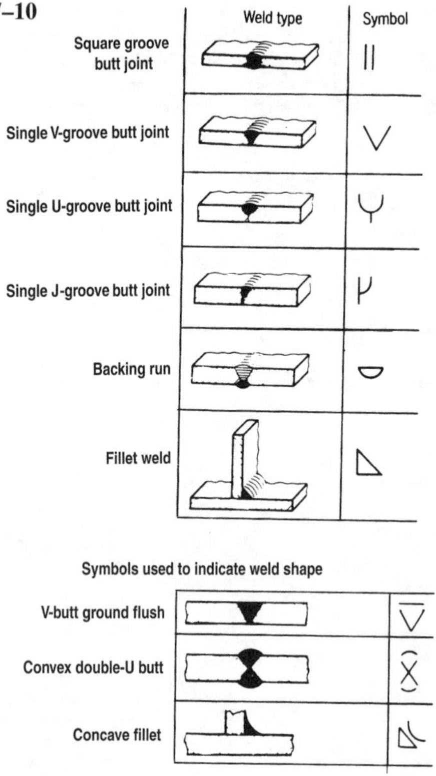

	Weld type	Symbol
Square groove butt joint		
Single V-groove butt joint		
Single U-groove butt joint		
Single J-groove butt joint		
Backing run		
Fillet weld		

Symbols used to indicate weld shape

V-butt ground flush		
Convex double-U butt		
Concave fillet		

17.4 Welding Defects

All welding processes, particularly the manual ones, can suffer from defects. The causes of these are reasonably predictable (see Fig. 17.11). Many weld defects can be detected by close visual inspection backed up by surface non-destructive examination (NDE).

Fig. 17–11

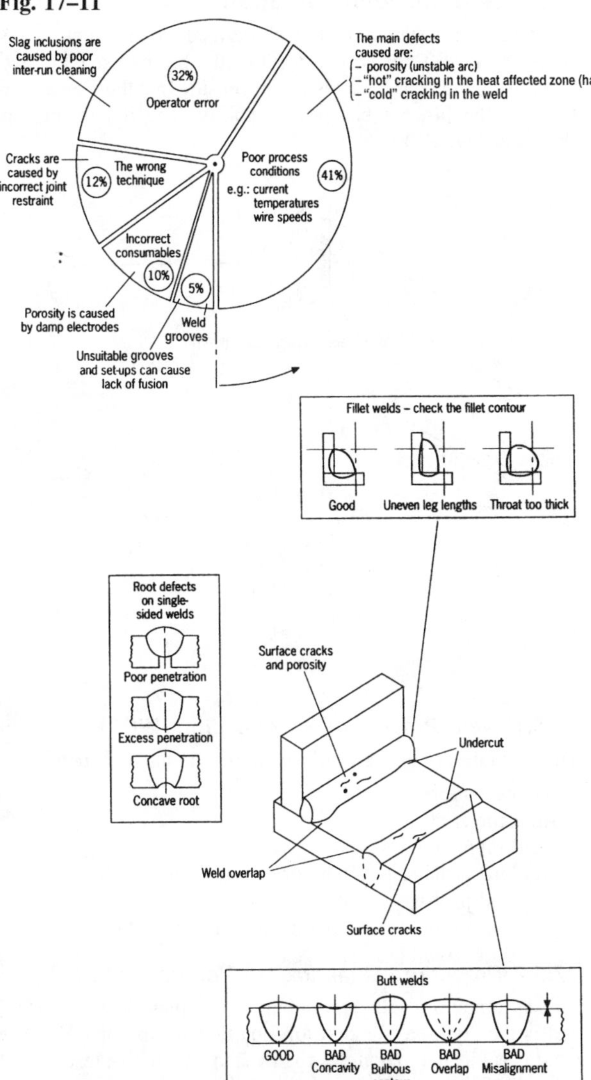
Slag inclusions are caused by poor inter-run cleaning
32%
Operator error
The main defects caused are:
– porosity (unstable arc)
– "hot" cracking in the heat affected zone (haz)
– "cold" cracking in the weld
Cracks are caused by incorrect joint restraint
12%
The wrong technique
Poor process conditions
41%
e.g.: current temperatures wire speeds
Incorrect consumables
10%
5%
Porosity is caused by damp electrodes
Weld grooves
Unsuitable grooves and set-ups can cause lack of fusion
Fillet welds – check the fillet contour
Good
Uneven leg lengths
Throat too thick
Root defects on single-sided welds
Poor penetration
Excess penetration
Concave root
Surface cracks and porosity
Undercut
Weld overlap
Surface cracks
Butt welds
GOOD
BAD Concavity
BAD Bulbous contour
BAD Overlap
BAD Misalignment

17.5 Welding Documentation

Welding is associated with well-defined documentation designed to specify the correct weld method to be used, confirm that this method has been tested, and ensure that the welder performing the process has proven ability. The documents are shown in Fig. 17.12.

Fig. 17–12

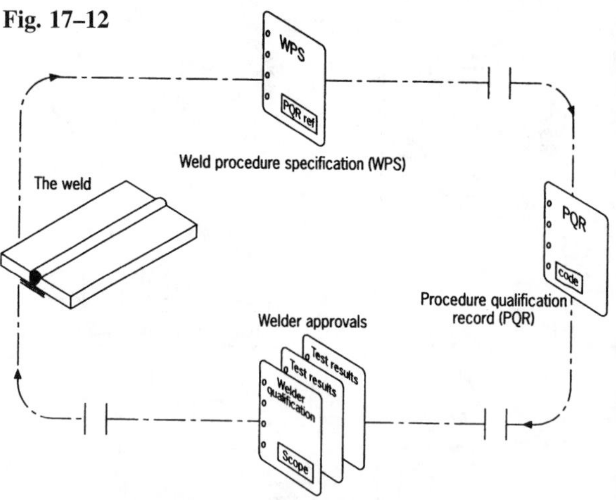

17.5.1 Weld Procedure Specification (WPS)

The WPS describes the weld technique and includes details of:

—parent material;
—filler material;
—weld preparation;
—welding variables; current, orientation, etc.;
—pre- and post-weld heat treatment; and
—the relevant procedure qualification record (PQR).

17.5.2 Procedure Qualification Record (PQR)

This is sometimes called a weld procedure qualification (WPQ) and is the "type-test" record of a particular type of weld. The weld is subjected to non-destructive and destructive tests to test its quality.

17.5.3 Welder Qualifications

"Coded" welders are tested to a range of specific WPSs to ensure their technique is good enough.

17.6 Useful References

1. ASME Section IX: *Boiler and Pressure Vessel Code—Welding.*
2. AWS D1.1: 1996. *Structural Welding Code—Steel.*
   D1.2: 1990. *Structural Welding Code—Aluminum.*
   D1.3: 1989. *Structural Welding Code—Sheet Steel.*
3. AWS A2.4: 1993. *Standard Symbols for Welding Brazing and Non-Destructive Examination.*
4. AWS A5.01: 1993. *Filler Metal Procurement Guidelines.*
5. ISO 14732: 1998. *Approval Testing of Welding Personnel.*
6. ISO 9692: 1992. *Metal Arc Welding.*
7. AWS B2.1-1-206: 1996. *Shielded Metal Arc Welding.*

Reference Books

1. *Technical Data Handbook*: 1986: Pub: The Hartford Steam Boiler Inspection and Insurance Co.
2. Publications of the Lincoln Electric Co. Inc including: *The Procedure Handbook of Arc Welding:* 13th edition.

Web Sites

Welding Research Council: www.forengineers.org/wrc/index.htm
The American Welding Society: www.amweld.org

Section 18

Non-Destructive Examination (NDE)

18.1 NDE Acronyms

Non-destructive examination procedures, reports, and general literature are full of acronyms. The most common ones are shown in Table 18.1.

Table 18.1 NDE Acronyms

AE	Acoustic Emission
AFD	Automated Flaw Detection
A-Scan	Amplitude Scan
ASNT	American Society for Non-Destructive Testing
ASTM	American Society for Testing and Materials
B-Scan	Brightness Scan
BVID	Barely Visible Impact Damage
CDI	Crack Detection Index
CRT	Cathode Ray Tube
C-Scan	Contrast Scan
CSI	Compton Scatter Imaging
CTM	Coating Thickness Measurement
CW	Continuous Wave/Compression Wave
DAC	Distance Amplitude Correction
dB	Decibel
DGS	Distance, Gain, Size (Diagram)
DPEC	Deep Penetration Eddy Currents
EC	Eddy Current
ECII	Eddy Current Impedance Imaging
EPS	Equivalent Penetrameter Sensitivity
ET (ECT)	Eddy Current Testing
FFD	Focus-to-Film Distance
FSH	Full Scale Height

Table 18.1 NDE Acronyms (cont.)

HAZ	Heat Affected Zone
HDR	High Definition Radiography
HVT	Half Value Thickness
IF	Industrial Fiberscope
IQI	Image Quality Indicator
IV	Industrial Videoimagescope
LD	Linear Detectors
LFECA	Low Frequency Eddy Current Array
LPI	Liquid Penetrant Inspection
LW	Longitudinal Wave
MFL	Magnetic Flux Leakage
MPI	Magnetic Particle Inspection
MPT	Magnetic Particle Testing
MR	Microradiography
MRI	Magnetic Resonance Imaging
MT	Magnetic Testing
NDA	Non-Destructive Assessment
NDE	Non-Destructive Examination
NDI	Non-Destructive Inspection
NDT	Non-Destructive Testing
NMR	Nuclear Magnetic Resonance
PA	Peak Amplitude
PDRAM	Pulsed Digital Reflection Acoustic Microscopy
POD	Probability of Detection
P-Scan	Projection Scan
PT	Penetrant Testing
PVT	Pulse Video Thermography
QNDE	Quantitative Non-Destructive Evaluation
RFECT	Remote Field Eddy Current Testing
ROI	Region of Interest
ROV	Remotely-Operated Vehicle
RT	Radiographic Testing
RT	Real Time
RTUIS	Real Time Ultrasonic Imaging System
RVI	Remote Visual Inspection
RVT	Remote Visual Testing

Table 18.1 NDE Acronyms (cont.)

SAM	Scanning Acoustic Microscopy
SDT	Static Deflection Techniques
SEM	Scanning Electron Microscopy
SFD	Source-to-Film Distance
SH	Horizontally Polarized Shear Waves
SI	Sensitivity Indicator
SIT	Simulated Infrared Thermography
SMNR	Signal-to-Material Noise Ratio
SNR	Signal-to-Noise Ratio
SPATE	Stress Pattern Analysis by Thermal Emission
TDR	Time-Domain Reflectometry
TOFD	Time-of-Flight Diffraction
TSE	Total Spectral Energy
TW	Transverse Wave
US	Ultrasonic
UT	Ultrasonic Testing
VAP	Variable Angle (Ultrasonic) Probe
VT	Visual Testing
WFMPI	Wet Fluorescent Magnetic Particle Inspection
WIR	Work and Inspection Robot
WT	Wall Thickness

NDE techniques are in common use to check the integrity of engineering materials and components. The main applications are plate, forgings, castings, and welds.

18.2 Visual Examination

Close visual examination can reveal surface cracks and defects of about "40 mils" or 0.004 in (0.1 mm) and above. This is larger than the "critical crack size" for most ferrous materials.

18.3 Dye Penetrant (DP) Testing

This is an enhanced visual technique using three aerosols, a cleaner (clear), penetrant (red), and developer (white). Surface defects appear as a thin red line (see Fig. 18.1).

Fig. 18–1

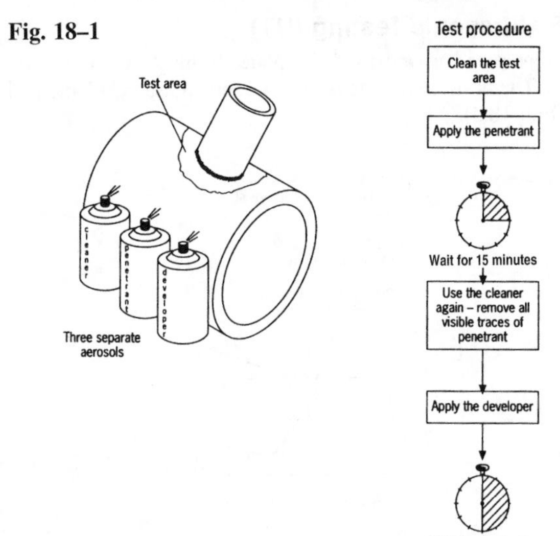
Test area
cleaner
penetrant
developer
Three separate aerosols
Test procedure
Clean the test area
Apply the penetrant
Wait for 15 minutes
Use the cleaner again – remove all visible traces of penetrant
Apply the developer
Wait 30 minutes for any indications to "develop"

18.4 Magnetic Particle (MP) Testing

This works by passing a magnetic flux through the material while spraying the surface with magnetic ink. An air gap in a surface defect forms a discontinuity in the field, which attracts the ink, making the crack visible (see Fig. 18.2).

Fig. 18–2

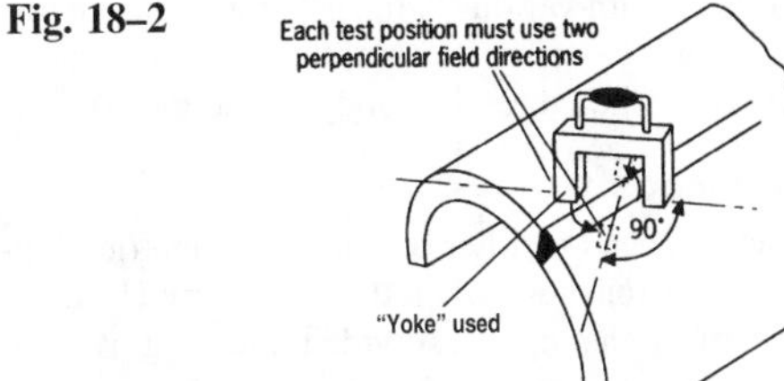
Each test position must use two perpendicular field directions
90°
"Yoke" used

18.5 Ultrasonic Testing (UT)

Different practices are used for plate, forgings, castings, and welds. The basic technique is the "A-scope pulse-echo" method (see Fig. 18.3).

Fig. 18–3

- A "pulsed" wave is used–it reflects from the back wall, and any defects.
- The location of the defect can be read off the screen.

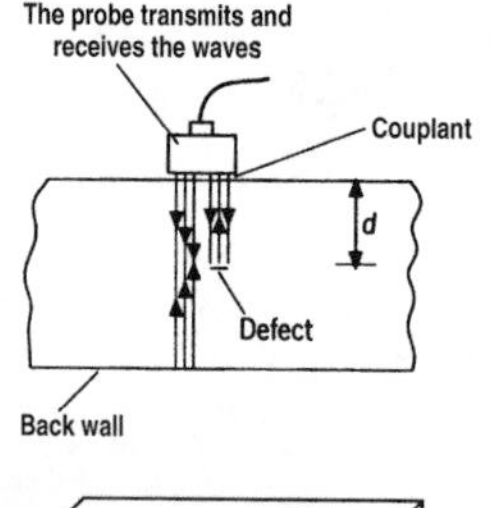

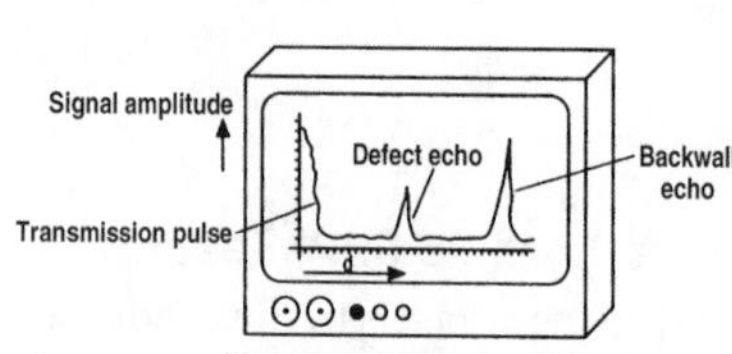

The horizontal axis represents time–i.e., the "distance" into the material

18.5.1 UT of Plate

Technical standards contain various "grades" of acceptance criteria. The plate is tested to verify its compliance with a particular grade specified for the edges and body of the material.

18.5.2 UT of Castings

Casting discontinuities can be either planar or volumetric. Separate gradings are used for these when discovered by UT technique. The areas of a casting are divided into critical and non-critical areas, and by thickness "zones" (see Fig. 18.4).

Fig. 18–4

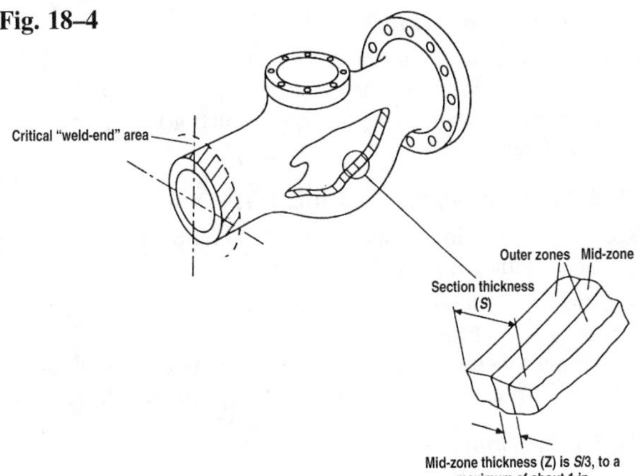

18.5.3 UT of Welds

Weld UT has to be a well-controlled procedure because the defects are small and difficult to classify. Ultrasonic scans may be necessary from several different directions, depending on the weld type and orientation (see Fig. 18.5).

Fig. 18–5

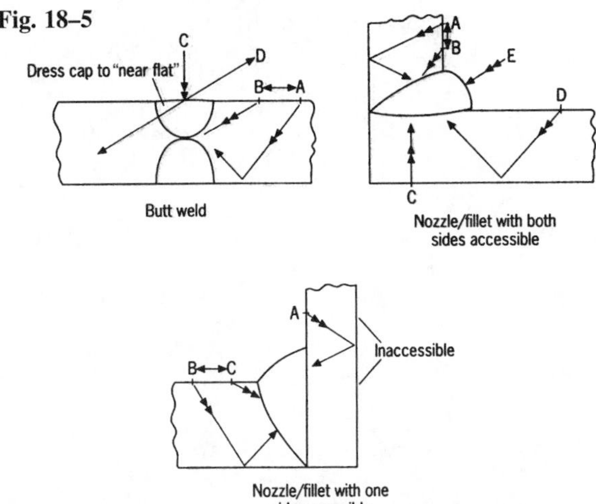

The general technique is:

—Surface scan using normal (0°) probe.
—Transverse scan (across the weld) to detect *longitudinal* defects.
—Longitudinal scan (along the weld direction) to detect *transverse* defects.

18.6 Radiographic Testing (RT)

Radiography is widely used for NDT of components and welds in many engineering applications.

—X-rays are effective on steel up to a thickness of approximately 6 in (150 mm).
—Gamma (γ) rays can also be used for thickness of 2-6 in (50–150 mm) but definition is not as good as with X-rays.

18.6.1 Techniques

For tubular components, a single or double wall technique may be used. Figure 18.6 shows the technique.

Fig. 18–6

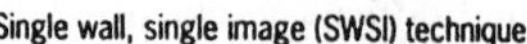

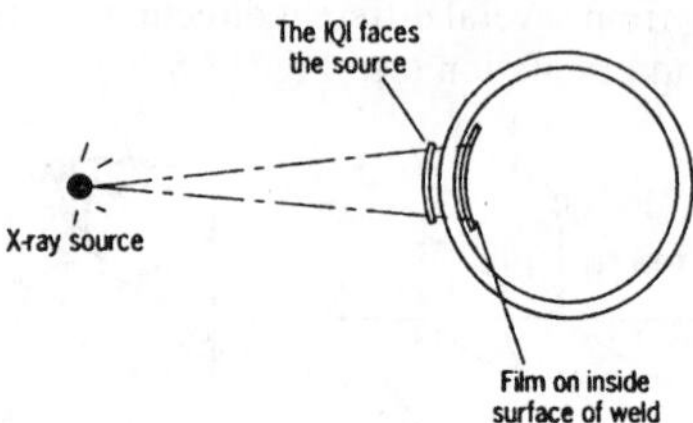

Fig. 18–6 (cont.)

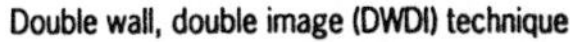
Double wall, double image (DWDI) technique

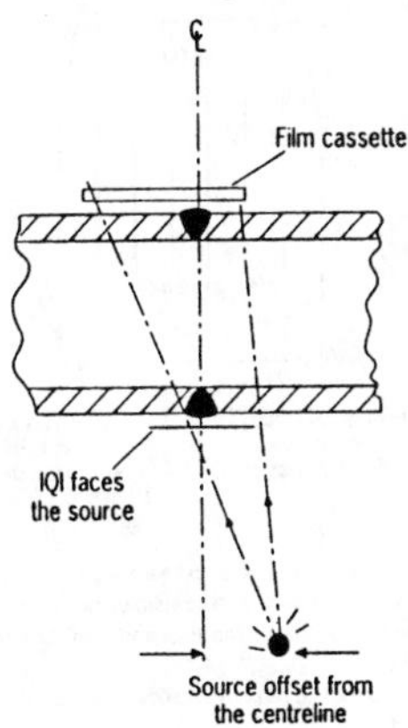

18.6.2 Penetrameters

Penetrameters, or image quality indicators (IQIs), check the sensitivity of a radiographic technique—to ensure that any defects present will be visible. The two main types are the "wire" type and "hole" type (see Fig. 18.7).

Fig. 18–7

Wire type IQI

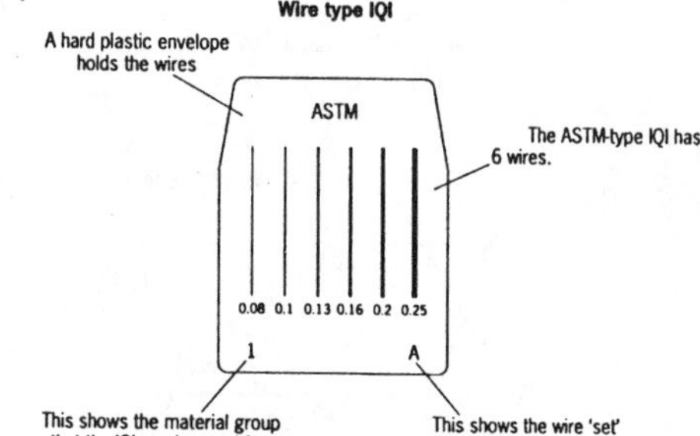
A hard plastic envelope holds the wires
ASTM
The ASTM-type IQI has 6 wires.
0.08 0.1 0.13 0.16 0.2 0.25
1
A
This shows the material group that the IQI can be used for. Group 1 is carbon/alloy/stainless steels
This shows the wire 'set' size. In this metric example the 'set A' wire sizes are: 0.08 mm, 0.1 mm, 0.13 mm, 0.16 mm, 0.2 mm, and 0.25 mm. Sets B,C,D have larger wire diameters

- The objective is to look for the smallest wire visible
- Sensitivity = diameter of smallest wire visible/maximum penetrated thickness of weld
- If the above IQI is used on 10mm material and the 0.16mm wire is visible, then sensitivity = 0.16/10 = 1.6%
- Check the standard for the maximum allowable sensitivity for the technique/application being used

Hole type IQI

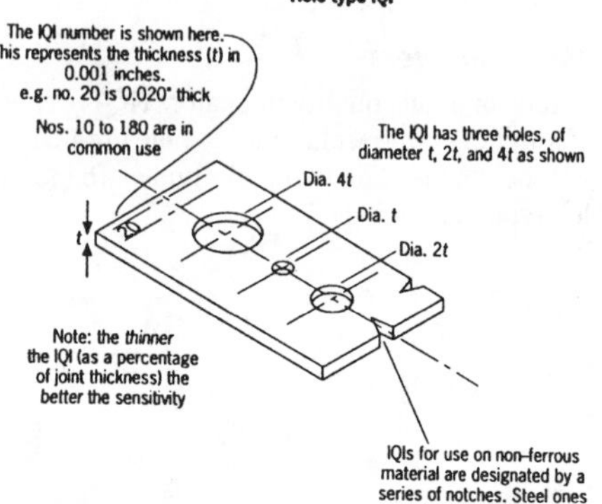
The IQI number is shown here. This represents the thickness (t) in 0.001 inches. e.g. no. 20 is 0.020" thick
Nos. 10 to 180 are in common use
The IQI has three holes, of diameter t, 2t, and 4t as shown
Dia. 4t
Dia. t
Dia. 2t
20
t
Note: the thinner the IQI (as a percentage of joint thickness) the better the sensitivity
IQIs for use on non-ferrous material are designated by a series of notches. Steel ones have no notches.

IQI designation	*Sensitivity*	*Visible hole**
1–2t	1	2t
2–1t	1.4	1t
2–2t	2.0	2t
2–4t	2.8	4t
4–2t	4.0	2t

* The hole that must be visible in order to ensure the sensitivity level shown

18.7 Useful References

Mix. P. E. *Introduction to Non Destructive Testing—A Training Guide*: 1987: Pub: Wiley. ISBN 0-471-8312G-3.

Standards: Visual Examination

1. ASTM A802/A802M: 1996. *Visual Examination of Steel Castings.*
2. MSS-SP-55: 1984. *Quality Standard for Steel Castings.*
3. EN 12062:1998. *Non Destructive Examination of Welds—General Rules for Metallic Materials.*

Standards: Dye Penetrant Testing

1. ASTM E 165: 1982. *Dye Penetrant Examination.*
2. EN 571-1:1997. *Penetrant Testing—General Principles.*

Standards: Magnetic Particle Testing

1. ASTM E 1444: 1984. *Magnetic Particle Detection for Ferromagnetic Materials.*
2. EN 1290:1998. *Magnetic Particle Examination of Welds.*

Standards: UT Testing of Plate

1. ASTM A945/A945M: 1995. *Standard Specification for Steel Plate.*
2. EN 160: 1993: *Specification for Acceptance Levels for Internal Imperfections In Steel Plate, Based on Ultrasonic Testing.*

Standards: UT Testing of Castings

1. MIL-HDBK-728/6 NOT 1: 1992. *Ultrasonic Testing.*
2. ASTM A609: 1991. *Practice for Ultrasonic Examination of Castings.*

Standards: Radiography Techniques

1. ASTM E142: 1992. *Standard Methods for Controlling the Quality of Radiographic Testing.*
2. MIL-HDBK-728/5A: 1994. *Radiographic Testing.*
3. ASTM E04: 1993. *Guide for Radiographic Testing.*

Web Sites

American Society of Non-Destructive Testing: www.asnt.org
ASNT Research Journal: www.msc.cornell.edu/~nde
NASA NDE Branch: http//nesb.larc.nasa.gov
NDE Newsgroups home page: www.nde.swri.edu
ASNT NDE Handbook home page: www.asnt.org/publications/handbook.htm

Section 19

Corrosion

19.1 Types of Corrosion

There are three basic types of corrosion:

- Chemical corrosion
- Galvanic corrosion
- Electrolytic corrosion

To complicate matters there are a variety of sub-types, some hybrids, and a few that do not fit neatly into any single category.

19.1.1 Chemical Corrosion

This is caused by attack by chemical compounds in a material's environment. It is sometimes referred to as "dry" corrosion or oxidation.

Examples are:

- Oxidation (scaling) of iron at high temperatures.
- Oxidation of Ni in sulphurous gas.

19.1.2 Galvanic Corrosion

This is caused by two or more dissimilar metals in contact in the presence of a conducting electrolyte. One material becomes anodic to the other and corrodes (see Fig. 19.1). Examples are:

- Stainless steel trim causes anodic corrosion on carbon-steel automobile bodywork.
- Defective coating of tin on carbon steel increases the corrosion rate of steel.

Fig. 19–1

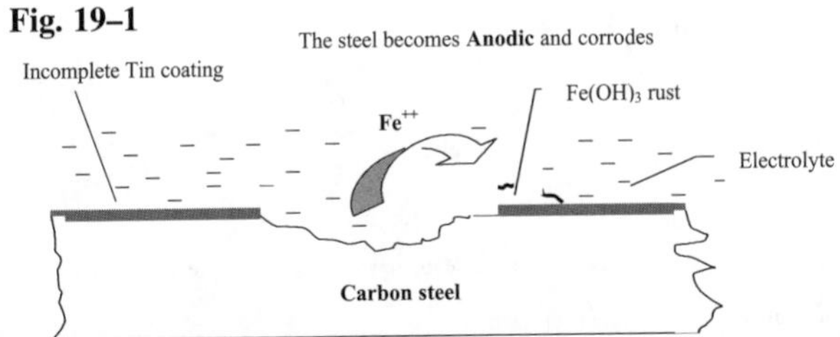

The tendency of a metal to become anodic or cathodic is governed by its position in the electrochemical series (see Fig. 19.2). This is, strictly, accurate only for pure metals rather than metallic compounds and alloys (see ASTM G135 and ASTM G102). A more general guide to galvanic corrosion attack of common engineering materials is given in Fig. 19.3.

Fig. 19–2

THE ELECTROCHEMICAL SERIES

Galvanic corrosion occurs when dissimilar metals are in contact with a conducting electrolyte. The electrochemical series shows the relative potentials of pure metals.

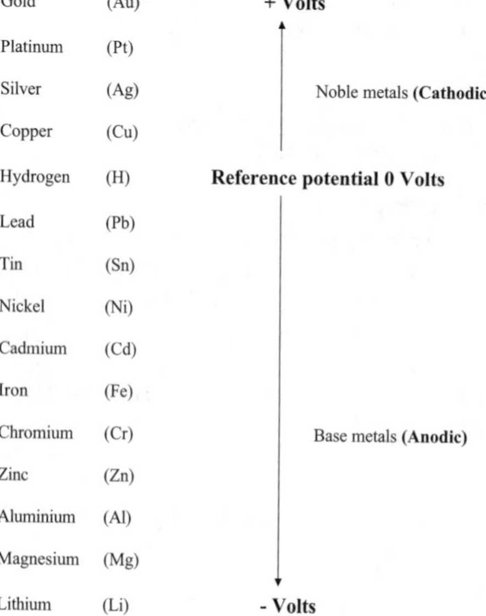

Remember: Metals higher in the table become cathodic and are protected by the (anodic) metals below them in the table.

Fig. 19–3

GALVANIC CORROSION ATTACK ¾ GUIDELINES

Corrosion of the materials in each column is increased by contact with the materials in the row when the corresponding box is shaded.

Material	Steel and CI	Stainless steel 18% Cr	Stainless steel 11% Cr	Inconel Ni alloys	Cu/Ni and bronzes	Cu and brass	PbSn and soft solder	Silver solder	Mg alloys	Chromium	Titanium	Al alloys	Zinc
Steel and CI		■	■	■	■	■	■	■		■	■	■	
Stainless steel 18% Cr													
Stainless steel 11% Cr		■		■	■	■		■		■	■		
Inconel/Ni alloys													
Cu/Ni and bronzes		■		■						■	■		
Cu and brass		■	■	■	■		■	■		■	■		
PbSn and soft solder		■	■	■	■	■		■		■	■		■
Silver solder		■		■						■	■		
Mg alloys	■	■	■	■	■	■	■	■		■	■	■	■
Chromium													
Titanium													
Al alloys	■	■	■	■	■	■	■	■		■	■		
Zinc	■	■	■	■	■	■	■	■		■	■	■	

Example: The corrosion rate of silver solder is increased when it is placed in contact with 11% Cr stainless steel.

19.1.3 Electrolytic Corrosion

This is sometimes referred to as "wet" or "electrolytic" corrosion. It is similar to galvanic corrosion in that it involves a potential difference and an electrolyte but it does not have to have dissimilar materials. The galvanic action often happens on a microscopic scale. Examples are:

- Pitting of castings due to galvanic action between different parts of the crystals (which have different compositions).
- Corrosion of castings due to grain boundary corrosion (see Fig. 19.4).

Fig. 19–4

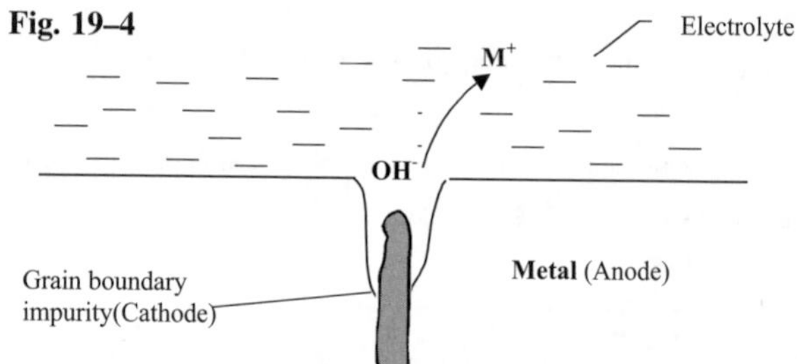

19.2 Crevice Corrosion

This occurs between close-fitting surfaces, crevice faces, or anywhere where a metal is restricted from forming a protective oxide layer (see Fig. 19.5). Corrosion normally propagates in the form of pitting. Examples are:

- Corrosion in crevices in seal welds.
- Corrosion in lap joints used in fabricated components and vessels.

Fig. 19–5

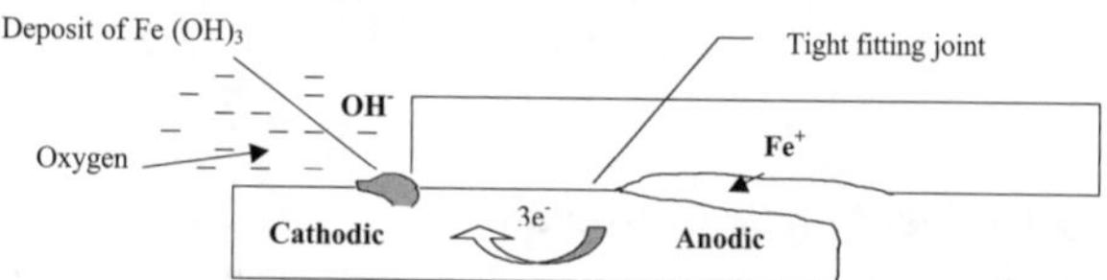

19.3 Stress Corrosion

This is caused by a combination of corrosive environment and tensile loading. Cracks in a material's brittle surface layer propagate into the material, resulting in multiple bifurcated (branching) cracks. Examples are:

- Failure in stainless steel pipes and bellows in a chlorate-rich environment.
- Corrosion of austenitic stainless steel pressure vessels.

19.4 Corrosion Fatigue

This is a hybrid category in which the effect of a corrosion mechanism is increased by the existence of a fatigue condition. Seawater, fresh water, and even air can reduce the fatigue life of a material.

19.5 Intergranular Corrosion

This is a form of local anodic attack at the grain boundaries of crystals due to microscopic difference in the metal structure and composition. Examples are:

- "Weld decay" in unstabilized austenitic stainless steels.
- "Dezincification" of brass in seawater, the selective removal of zinc from the alloy leaving behind a spongy mass of copper.

19.6 Erosion–Corrosion

Almost any corrosion mechanism is made worse if the material is subject to simultaneous corrosion and abrasion. Abrasion removes the protective passive film that forms on the surface of many metals, exposing the underlying metal. An example is:

- The walls of pipes containing fast-flowing fluids and suspended solids.

19.7 Useful References

Web Sites

National Association of Corrosion Engineers (NACE): www.nace.org

For a list of corrosion-related links: www.nace.org/corlink/corplinkindex.htm

Standards

ASTM G15: 1999: *Standard Terminology Relating to Corrosion and Corrosion Testing.*

ASTM G135: 1995: *Standard Guide for Computerized Exchange of Corrosion Data for Metals.*

ASTM G119: 1998: *Standard Guide for Determining Synergism Between Wear and Corrosion.*

ASTM G102: 1999: *Standard Practice for Calculation of Corrosion Rates and Related Information From Electrochemical Measurements.*

Section 20

Surface Protection

20.1 Painting

There are numerous types of paint and application techniques. Correct preparation, choice of paint system, and application are necessary if the coating is to have the necessary protective effect.

20.1.1 Preparation

Commonly used surface preparation grades are taken from the ASTM Steel Structures Painting Council (SSPC) and the Swedish standard SIS O5 5900 (see Table 20.1)

Table 20.1 Surface Preparation Grades

SSPC Designation	*Preparation Grade*	*SIS 05 5900 Designation*
SSPC SP5	Blast cleaning to pure metal. No surface staining remaining.	Sa 3
SSPC SP10	Thorough blast cleaning but some surface staining may remain.	Sa 2½
SSPC SP6	Blast cleaning to remove most of the millscale and rust.	Sa 2
SSPC SP7	Light blast cleaning to remove the worst millscale and rust.	Sa 1

20.1.2 Paint Types

These are divided broadly into air-drying, two-pack, and primers.

—Air-drying types: alkyd resins, esters, and chlorinated rubbers.

—Two-pack types: epoxy, polyurethanes.

—Primers: zinc phosphate or zinc chromate.

20.1.3 Typical Paint System

Most paint systems for outdoor use have a minimum of three coats, with a final dry film thickness (dft) of 6000–7500 micro inch (152–190 μm) (see Fig. 20.1).

Fig. 20–1

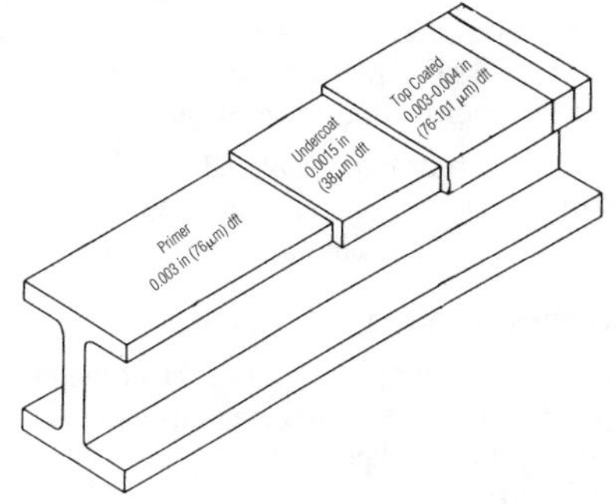

20.2 Galvanizing

Galvanizing is the generic term for the coating of iron and steel components with zinc. It can be used instead of painting to protect the base material from corrosion. The coating is usually applied by weight, in accordance with SSPC recommendations. Guidelines are shown in Table 20.2.

Table 20.2 Galvanizing Weights

Base Material	*Minimum Galvanized Coating Weight (oz/ft^2)*
Steel up to 1⁄16 in thick	1.1 (335 g/m^2)
Steel 1⁄16 to 1⁄4 in thick	1.5 (460 g/m^2)
Steel > 1⁄4 in thick	2 (610 g/m^2)
Castings	2 (610 g/m^2)

An approximate conversion from coating weight to coating thickness is:

$$1\ g/m^2 \cong 0.14\ \mu m;\ 1\ oz/ft^2\ (304.5\ g/m^2) \cong 1679 \text{ micro inch } (42.6\ \mu m)$$

Coating uniformity is tested by a "Preece test," which involves exposing a coated specimen to a salt solution.

20.3 Chrome Plating

1. Chrome plating provides a fine finish for hydraulic components and provides protection against some environmental conditions. The process is well covered by technical standards such as SAE-AMS 2406H: *Chromium Plating.*

   A typical specification for a plated component is:

   - 800 micro inch of copper plated onto the steel.
   - 1000 micro inch of nickel plated onto the copper.
   - 20 micro inch of chrome as a top layer.

20.4 Rubber Lining

Rubber lining is commonly used to protect materials in seawater and chemical process systems against corrosive and erosive attack. It is applied in sheets up to about ¼ inch thick. There are two main types:

- **Natural rubbers**—for low temperatures in oil-free water or slurry applications.
- **Synthetic rubbers**—(nitryl, butyl, neoprene)—for temperatures up to 240°F, or when oil is present.

20.4.1 Rubber Properties

Both natural and synthetic rubbers can be divided into hard and soft types. Two hardness scales are in use: IRHD (international rubber hardness degrees) and the "Shore" scale.

- **Hard** rubber (sometimes called ebonite) is 80–100 degrees IRHD or 60–80 "Shore D" scale.
- **Soft** rubber is 40–80 degrees IRHD or 40-80 "Shore A."

20.4.2 Design Features

Rubber-lined components need to have specific design features to help the lining adhere properly (see Fig. 20.2).

Fig. 20–2

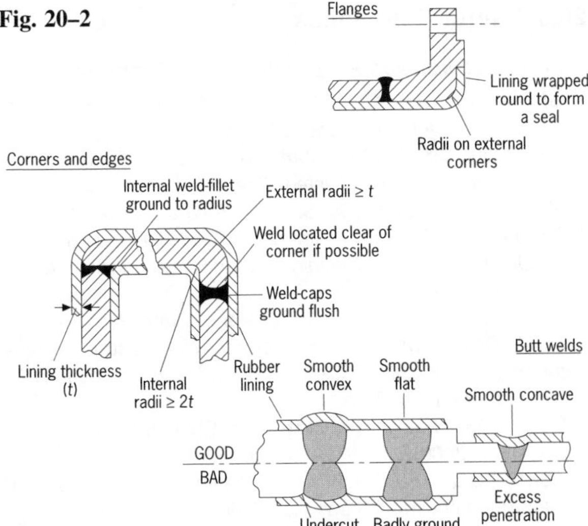

20.4.3 Application

The basic application procedure is:

1. Shotblast the metal surface to grade SSPC 10.
2. Apply adhesive to the surface.
3. Lay the sheets of rubber manually in scarf-jointed overlapping courses.
4. Vulcanize the rubber by heating to approximately 250°F (120°C) using steam or hot water.

20.4.4 Testing

Common tests on the applied rubber lining are:

- Spark testing (≅ 20 kV)—to check the continuity of the lining
- Rapping test—using a special hammer to test the adhesion of the lining to the metal.
- Hardness test—using a hand-held gage to measure the Shore or IRHD hardness. This shows whether vulcanization is complete.

20.5 Useful References

Standards: Surface Preparation and Coatings

1. SSPC Doc VIS 1: 1989. *Visual Standard for Abrasive Blast Cleaning of Steel. (reference photographs).*
2. SSPC Doc VIS 2: 1982. *Standard Method of Evaluating the Degree of Rusting on Painted Steel Surfaces.*
3. SSPC Doc SPCOM: 1995. *Surface Preparation Specifications.*
4. SIS 05 5900: *Pictorial Standards for Blast-Cleaned Steel (and for Other Methods of Cleaning).* (Standardiseringskommissionen) I Sverige, Stockholm.
5. SSPC: *Steel Structures Painting Manual Vol 1.*
6. ASTM D1186: 1993. *Test Methods for Non-Destructive Measurement of Dry Film Thickness of Non-Magnetic Coatings Applied to a Ferrous Base.*
7. SSPC: *Good Painting Practice Vol 1*: Chapter 21.0.
8. SAE-AMS 2406H: 1994. *Chromium Plating.*
9. ASSIST QQ-C-320B(4): 1987: *Chromium Plating.* This document references standards ASTM B504 and B556.

Standards: Rubber Linings

1. KS Document No M6691: 1978. *Recommended Practice for Rubber Lining of Vessels and Equipment.*
2. DIN (Germany) 28 051: 1990. *Chemical Apparatus—Designs of Metal Components to be Protected by Organic Coatings or Linings.*

Section 21

Metallurgical Terms

Terminology used in metallurgy is complex. Some of the more common (and sometimes misunderstood) terms are given below:

age hardening Hardening by aging, usually after rapid cooling or cold working.

aging A change of properties that occurs at ambient or moderately elevated temperatures after hot working, heat treating, quenching, or cold working.

alloy A substance having metallic properties and composed of two or more chemical elements of which at least one is a metal.

alloy steel Steel containing significant quantities of alloying elements (other than carbon and small amounts of manganese, silicon, sulphur, and phosphorus) added to produce changes in mechanical or physical properties. Those containing less than about 5 percent total metallic alloying elements are termed low-alloy steels.

annealing Heating metal to a suitable temperature followed by cooling to produce discrete changes in microstructure and properties.

austenite A solid solution of one or more alloying elements in the FCC structure of iron.

bainite A eutectoid transformation product of ferrite and dispersed carbide.

beach marks Crack arrest "lines" seen on fatigue fracture surfaces.

billet A solid piece of steel that has been hot worked by forging, rolling, or extrusion.

brittle fracture Fracture preceded by little or no plastic deformation.

brittleness The tendency of a material to fracture without first undergoing significant plastic deformation.

carbide A compound of carbon with metallic elements (e.g., tungsten, chromium).

carbon equivalent (CE) A "weldability" value that takes into account the effects of carbon and other alloying elements on a particular characteristic of steel. A formula commonly used is:

$$CE = C + (Mn/6) + [(Cr + Mo + V)/5] + [(Ni + Cu)/15]$$

carbon steel A steel containing only small quantities of elements other than carbon.

cast iron Iron containing more than about 2 percent carbon.

cast steel Steel castings, containing less than 2 percent carbon.

cementite A carbide, with composition Fe_3C.

cleavage Fracture of a crystal by crack propagation.

constitutional diagram A graph showing the temperature and composition limits of various phases in a metallic alloy.

crack initiator Physical feature that encourages a crack to start.

creep Time-dependent strain occurring under stress.

critical cooling rate The maximum rate at which austenite needs to be cooled to ensure that a particular type of structure is formed.

crystalline The general structure of many metals.

crystalline fracture A fracture of a metal showing a grainy appearance.

decarburization Loss of carbon from the surface of a ferrous alloy caused by heating.

deformation General term for strain or elongation of a metal's lattice structure.

duplex Containing two phases (e.g., ferrite and pearlite).

deoxidation Removal of oxygen from molten metals by use of chemical additives.

diffusion Movement of molecules through a solid solution.

dislocation A linear defect in the structure of a crystal.

ductility The capacity of a material to deform plastically without fracturing.

elastic limit The maximum stress to which a material may be subjected without any permanent deformation occurring.

equilibrium diagram A graph of the temperature, pressure, and composition limits of the various phases in an alloy "system."

etching Subjecting the surface of a metal to an acid to reveal the microstructure.

fatigue A cycle or fluctuating stress condition leading to fracture.

ferrite A solid solution of alloying elements in BCC iron.

fibrous fracture A fracture whose surface is characterized by a dull or silky appearance.

grain An individual crystal in a metal or alloy.

grain growth Increase in the size of the grains in metal caused by heating at high temperature.

graphitization Formation of graphite in iron or steel.

hardenability The property that determines the depth and distribution of hardness induced by quenching.

hardness (indentation) Resistance of a metal to plastic deformation by indentation (measured by Rockwell, Brinel, or Vickers test).

inclusion A metallic or non-metallic material in the matrix structure of a metal.

initiation point The point at which a crack starts.

killed steel Steel deoxidized with silicon or aluminium, to reduce the oxygen content.

K_{1C} A fracture toughness parameter.

lamellar tear A system of cracks or discontinuities, normally in a weld.

lattice A pattern (physical arrangement) of a metal's molecular structure.

macrograph A low-magnification picture of the prepared surface of a specimen.

macrostructure The structure of a metal as revealed by examination of the etched surface at a magnification of about × 15.

martensite A supersaturated solution of carbon in ferrite.

microstructure The structure of a prepared surface of a metal as revealed by a microscope at a magnification than about × 15.

micro-cracks Small "brittle" cracks, normally perpendicular to the main tensile axis.

necking Local reduction of the cross-sectional area of metal by stretching.

normalizing Heating a ferrous alloy and then cooling in still air.

notch brittleness A measure of the susceptibility of a material to brittle fracture at locations of stress concentration (notches, grooves, etc.).

notch sensitivity A measure of the reduction in strength of a metal caused by the presence of stress concentrations.

nitriding Surface hardening process using nitrogenous material.

pearlite A product of ferrite and cementite with a lamellar structure.

phase A portion of a material "system" that is homogenous.

plastic deformation Deformation that remains after release of the stress that caused it.

polymorphism The property whereby certain substances may exist in more than one crystalline form.

precipitation hardening Hardening by managing the structure of a material, to prevent the movement of dislocations.

quench hardening Hardening by heating and then quenching quickly, causing austenite to be transformed into martensite.

recovery Softening of cold-worked metals when heated.

segregation Non-uniform distribution of alloying elements, impurities, or phases in a material.

slip Plastic deformation by shear of one part of a crystal relative to another.

slip plane Plane of dislocation movement.

soaking Keeping metal at a predetermined temperature during heat treatment.

solid solution A solid crystalline phase containing two or more chemical species.

solution heat treatment Heat treatment in which an alloy is heated so that its constituents enter into solid solution and then are cooled rapidly enough to "freeze" the constituents in solution.

spheroidizing Heating and cooling to produce a spheroid or globular form of carbide in steel.

strain aging Aging induced by cold working.

strain hardening An increase in hardness and strength caused by plastic deformation at temperatures below the recrystallization range.

stress-corrosion cracking Failure by cracking under the combined action of corrosion and stress.

sulphur print A macrographic method of examining the distribution of sulphur compounds in a material (normally forgings).

tempering Supplementary heat treatment to reduce excessive hardness.

temper brittleness An increase in the ductile brittle transition temperature in steels.

toughness Capacity of a metal to absorb energy and deform plastically before fracturing.

transformation temperature The temperature at which a change in phase occurs.

transition temperature The temperature at which a metal starts to exhibit brittle behavior.

weldability Suitability of a metal for welding.

work hardening Hardening of a material due to straining or "cold working."

Useful References

Brown. C. D. *Dictionary of Metallurgy:* 1998. Pub: Wiley. ISBN 0-471-96155-8.

Section 22

Engineering Associations and Organizations: Contact Details

Acronym	*Organization*	*URL*	*Mailing Address*	*Telephone*	*Fax*
AA	The Aluminum Association Inc.	www.aluminum.org	900 19th St NW Washington, DC 20006	(202)-862-5100	(202)-862-5164
AAAI	American Associationfor Artificial Intelligence	www.aaai.org	445 Burgess Drive Menlo Park CA 94025-3442	(650)-328-3123	(650)-321-4457
AAAS	American Association for Advancement of Science	www.aaas.org	1200 New York Ave NW Washington, DC 20005	(202)-326-6400	
AAEE	American Academy of Environmental Engineers	www.enviro-engrs.org	130 Holiday Court Suite 100 Annapolis MD 21401	(410)-266-3311	(410)-266-7653
AAES	American Association of Engineering Societies	www.aaes.org	1111, 19th St NW Suite 403 Washington DC 20036	(202)-296-2237	(202)-296-1151

ACCA	Air Conditioning Contractors of America	www.acca.org	1712 New Hampshire Ave Washington DC 20009	(202)-483-9370	(202)-588-1217
ACEC	American Consulting Engineers Council	www.acec.org	1015, 15th St NW #802 Washington DC 20005	(202)-347-7474	(202)-898-0068
ACII	American Concrete Institute International	www.aci-int.org	PO Box 9094 Funnington Hills MI 48333	(248)-848-3700	(248)-848-3701
ACIL	American Council of Independent Laboratories	www.aci/org.	1629 K Street NW Washington DC 20006	(202)-887-5872	(202)-887-0021
ACM	Association for Computing Machinery	www.acm.org	1, Aston Plaza 1515 Broadway New York, NY 10036	(212)-869-7440	(212)-944-1318
ACS	American Ceramic Society	www.acers.org	PO Box 6136 Westerville Ohio 43086-6136	(614)-890-4700	(614)-899-6109
AES	Abrasive Engineering Society	www.nauticom.net. www/grind	PO Box 3157 Butler PA 16003	(724)-282-6210	(724)-282-6210
AGMA	American Gear Manufacturers Association	www.agma.org	1500 King St Suite 20 Alexandria VA 22314-2730	(703)-684-0211	(703)-684-0242

Acronym	*Organization*	*URL*	*Mailing Address*	*Telephone*	*Fax*
AIAA	American Institute of Aeronautics and Astronautics	www.aiaa.org	Suite 500 1801 Alexander Bell Drive Reston, VA 20191-4400	(703)-264-7500	(703)-264-7551
AICHE	American Institute of Chemical Engineers	www.aiche.org	Three Park Ave New York, NY 10016-5991	(212)-591-8100	(212)-591-889
AIE	American Institute of Engineers	www.members-aie.org	1018 Aopian Way El Sobrante CA 94803	(510)-223-8911	(510)-223-8911
AIME	American Institute of Mining, Metallurgical and Petroleum Engineers	www.aim.eng@ aimeng.org	Three Park Ave New York, NY 10016-5998	(212)-419-7676	(212)-371-9622
AIP	American Institute of Physics	www.aip.org.	One, Physics Ellipse College Park MD 20740-3843	(301)-209-3100	(301)-209-0843
AISI	American Iron and Steel Institute	www.steel.org	1101 17th SE NW Suite 1300 Washington DC 20036	(202)-452-7100	(202)-463-6573

AISC	American Institute of Steel Construction Inc.	www.aisc.org	One, E Wacker Drive Suite 3100 Chicago IL 60601-2001	(312)-670-2400	(312)-670-5403
AISE	Association of Iron and Steel Engineers	www.aise.org	3, Gateway Center Suite 1900 Pittsburgh PA 15222-1004	(412)-281-6323	(412)-281-4657
AMT	The Association for Manufacturing Technology	www.mfgtech.org	7901 Westpark Drive McLean VA 22102	(703)-893-2900	(703)-893-1151
ANS	American Nuclear Society	www.ans.org	555 N. Kensington Ave La Grange Park IL 60526	(708)-352-6611	(708)-352-0499
ANSI	American National Standards Institute	www.ansi.org	11, W. 42nd St New York NY 10036	(212)-642-4900	(212)-398-0023
API	American Petroleum Institute	www.api.org	1220 L St NW Washington DC 20005	(202)-682-8000	(202)-682-8232
APQC	American Productivity and Quality Center	www.apqc.org	123 North Post Oak Lane 3rd Floor Houston TX 77024	(713)-681-4020	(713)-681-8578

Acronym	*Organization*	*URL*	*Mailing Address*	*Telephone*	*Fax*
APWA	American Public Works Association	www.pubworks.org	1301 Pennsylvania Ave NW Suite 501 Washington DC 20004	(202)-393-2792	(202)-737-9153
ASA	American Statistical Association	www.amstat.org	1429 Duke St Alexandria VA 22314-3402	(703)-684-1221	(703)-684-2037
ASA	Acoustical Society of America	www.asastds@aip.org	120 Wall St 32nd Floor New York NY 1005-3993	(212)-248-0373	(212)-248-0146
ASAE	American Society of Agricultural Engineers	www.asae.org	2950 Niles Rd St Joseph MI 49085-9659	(616)-429-0300	(616)-429-3852
ASCE	American Society of Civil Engineers	www.asce.org	1801 Alexander Bell Drive Reston VA 20191-4400	(703)-295-6300	(703)-295-6222
ASEE	American Society for Engineering Education	www.asee.org	1818 N. St NW Suite 600 Washington DC 20036-2479	(202)-331-350	(202)-265-8504

ASHE	American Society for Healthcare Engineering	www.ashe.org	1, N Franklin 27th Floor Chicago IL 60600	(312)-422-3800	
ASHRAE	American Society of Heating, Refrigeration and Air Conditioning Engineers	www.ashrae.org	1791 Tullie Circle NE Atlanta GA 30329	(404)-636-8400	(404)-321-5478
ASME	American Society of Mechanical Engineers	www.asme.org	Three, Park Ave New York NY 10016-5990	(973)-882-1167	(973)-882-1717
ASMI	American Society for Metals International	www.asm-intl.org	9639 Kinsman Rd Materials Park OH 44073-0002	(440)-338-5151	(440)-338-4634
ASNT	American Society for Non-Destructive Testing	www.asnt.org	1711 Arlington Lane Columbus OH 43228-0518	(614)274-6003	(614)-274-6899
ASPE	American Society for Precision Engineering	www.aspe.net	PO Box 10826 Raleigh NC 27605-0826	(919)-839-8444	(919)-839-8039
ASQ	American Society for Quality Control	www.asq.org	161 West Wisconsin Ave Milwaukee WI 53203	(414)-272-8575	(414)-272-1734
ASSE	American Society of Safety Engineers	www.asse.org	1800 E. Oakton St Des Plaines IL 60018	(847)-699-2929	(847)-768-3434

Acronym	*Organization*	*URL*	*Mailing Address*	*Telephone*	*Fax*
ASTM	American Society for Testing of Materials	www.ansi.org	100, Barr Harbor Drive W. Conshohocken PA 19428-2959	(610)-832-9585	(610)-832-9555
AWS	American Welding Society	www.awweld.org	550 NW Le Jeune Rd Miami FL 33126	(305)-443-9353	(305)-443-7559
AWWA	American Water Works Association Inc.	www.awwa.org	6666 W Quincy Ave Denver CO 80235	(303)-794-7711	(303)-794-3951
FCI	Fluid Controls Institute Inc.	www.fluidcontrols institite.org	PO Box 1485 Pompano Beach FL 33061	(216)-241-7333	(216)-241-0105
FI	Forging Industry Association	www.forging.org	25 W Prospect Dve Suite 300 Cleveland OH 44115	(216)-781-6260	(216)-781-0102
FMG	Factory Mutual Global	www.fmglobal.com	Westwood Executive Center 100 Lowder Brook Drive Suite 1100 Westwood, MA 02090-1190	(781)-326-5500	(781)-326-6632
HTRI	Heat Transfer Research Inc.	www.htrinet.com	1500 Research Parkway Suite 100 College Station TX 77845	(409)-260-6200	(409)-260-6249

ICI	Investment Casting Institute	www.investmentcasting.org	8150 N Central Expressway Suite M1008 Dallas TX 75206-1602	(214)-368-8896	(214)-368-8852
ICOSE	International Council on Systems Engineering	www.icose.org	2150 N. 107th St Suite 205 Seattle WA 98133-9009	(206)-361-6607	(206)-367-8777
IEEE	Institute of Electrical and Electronic Engineers	www.ieee.org	445 Hoes Lane Piscataway NJ 08855-1331	(732)-981-0060	(732)-562-6388
IGTI	The International Gas Turbine Institute	www.asme.org/igti	ASME International 5775-B Glenridge Dv Suite 370 Atlanta GA 30328-5380	(404)-847-0072	(404)-847-0151
IIE	Institute of Industrial Engineers	www.iienct.org	25, Technology Park Norcross GA 30092	(700)-449-0460	(700)-441-3295
ISOPE	International Society for Optical Engineering	www.spie.org	PO Box 10 Bellingham WA 98227-0010	(360)-676-3290	(360)-647-1445
ITE	Institute of Transportation Engineers	www.ite.org	525 School St SW Suite 410 Washington DC 20024	(202)-554-8050	(202)-863-5486

Acronym	*Organization*	*URL*	*Mailing Address*	*Telephone*	*Fax*
MMMS	Minerals, Metals and Materials Society	www.tms.org	184 Thorn Hill Rd Warrendale PA 15086	(724)-776-9000	(724)-776-3770
MSS	Manufacturers Standardization Society of the Valve and Fittings Industry	www.mss-hq.com	127 Park Street NE Vienna VA 22180-4602	(703)-281-6613	(703)-281-6671
NACE	National Association of Corrosion Engineers	www.nace.org	1440 South Creek Drive Houston TX 7708	(281)-228-6200	(281)-228-6300
NADCA	North American Die Casting Association	www.diecasting.org	366 Madison Ave New York NY 10017	(847)-292-3600	(847)-292-3620
NAFE	National Academy of Forensic Engineers	www.nafe.org	174 Brady Ave Hawthorne NY 10532	(914)-741-0633	(914)-747-2988
NBBPVI	National Board of Boiler and Pressure Vessel Inspectors	www.nationalboard.org	1155 North High Street Columbus OH 43201	(614)-888-8320	(614)-888-0750
NFP	National Fire Protection Association	www.nfpa.org	1, Batterymarch Park PO Box 9101 Quincy MA 02269-9101	(617)-770-3000	(617)-770-0700

NFU	National Fluid Power Association	www.nfpa.com	3333 N Mayfair Rd Milwaukee WI 53222-3219	(414)-778-3344	(414)-778-3361
NIST	National Institute of Standards and Technology	www.nist.gov	100 Bureau Drive Gaithersburg MD 20899-0001	(301)-975-8205	(301)-926-1630
NSPE	National Society of Professional Engineers	www.nspe.org	1420 King Street Alexandria VA 22414-2794	(888)-285-6773	(703)-836-4875
PFI	Pipe Fabrication Institute	www.pfi-institute.org	655-32nd Ave, Suite 201 Lachine Qc.Canada HT8 3G6	(514)-634-3434	(514)-634-9736
PMI	Project Management Institute	http://www.pmi.org/pmihq/who.htm	4 Campus Blvd Newton Square PA 19073-3299	(610)-356-4600	(610)-356-4647
RAB	Registrar Accreditation Board	www.rabnet.com	PO Box 3003 Milwaukee WI 53201-3005	(888)-722-2440	(414)-765-8661
SAE	Society of Automotive Engineers	www.sae.org	400 Commonwealth Drive Warrendale PA 105096-001	(724)-776-4841	(724)-776-5760
SAME	Society of American Military Engineers	www.same.org	607 Prince St Alexandria VA 22314-3117	(800)-336-3097	(703)-684-0231

Acronym	*Organization*	*URL*	*Mailing Address*	*Telephone*	*Fax*
SEI	Structural Engineering Institute	www.seinstitute.org	1801 Alexander Bell Drive Reston VA 20191-4400	(703)-295-6360	(703)-295-6361
SES	Standards Engineering Society	www.ses.standards.org	13340 SW 96th Ave Miami FL 33176	(305)-971-4798	(305)-971-4799
SME	Society of Manufacturing Engineers	www.sme.org	One, SME Drive Dearborn MI 48121	(313)-271-1500	(313)-271-2861
SME	Society of Mining Engineers	www.smenet.org	8307 Shaffer Parkway Littleton CO 80127	(303)-973-9550	(303)-973-3845
SNAME	Society of Naval Architects and Marine Engineers	www.sname.org	601 Pavonia Ave Jersey City NJ 07306	(201)-798-4800	(201)-798-4975
SPE	Society of Plastics Engineers	www.4spe.org	PO Box 403 Brookfield CT 06804-0403	(203)-775-0471	(203)-775-8490
SPI	Society of The Plastics Industry Inc.	www.plastics industry.org	1801 K Street NW Suite 600K Washington DC 20006	(202)-974-5200	(202)-296-7005

SSPC	Steel Structures Painting Council	www.sspc.org	40 24th Street Pittsburgh PA 15222-4656	(412)-281-2331	(412)-281-9992
STI	Steel Tank Institute	www.steeltank.org	570 Oakwood Rd Lake Zurich IL 60047	(847)-438-8265	(847)-438-8766
STLE	Society of Tribologists and Lubrication Engineers	www.stle.org	840 Busse Highway Park Ridge IL 60068	(847)-825-5536	(847)-825-1456
TEMA	Tubular Exchanger Manufacturers Association Inc.	www.tema.org	25 N Broadway Tarrytown NY 10591	(914)-332-0040	(914)-332-1541
UFE	United Engineering Foundation	www.engfnd.org	Three Park Ave New York, NY 10016-5902	(212)-591-7836	(212)-591-7441
USMA	US Metric Association Inc.	http://lamar.colostateweb.edu/uhhillger/	10245 Andasol Ave North Ridge CA 91325-1504